New Economic Windows

Series Editor
MASSIMO SALZANO

Series Editorial Board

Arnab Chatterjee · Bikas K. Chakrabarti (Eds.)

Econophysics of Markets and Business Networks

Proceedings of the Econophys-Kolkata III

Springer

Arnab Chatterjee
Bikas K. Chakrabarti
Theoretical Condensed Matter Physics Division and
Centre for Applied Mathematics and Computational Science
Saha Institute of Nuclear Physics
Kolkata, India

ISBN 978-88-470-5568-1 Springer Milan Berlin Heidelberg New York

Springer is a part of Springer Science+Business Media
springer.com
© Springer-Verlag Italia 2007
Softcover re-print of the Hardcover 1st edition 2007

Cover design: Simona Colombo, Milano
Cover figure: © www.pixed2000.org
Typeset by the authors using a Springer Macro package
Data conversion: LE-TeX Jelonek, Schmidt & Vöckler GbR, Leipzig, Germany
Printing and binding: Grafiche Porpora, Segrate (MI)

Springer-Verlag Italia – Via Decembrio 28 – 20137 Milano

Printed on acid-free paper

Preface

Studies on various markets and networks by physicists are not very uncommon these days. Often the people from economics, finance and physics are not in agreement regarding the nature of the problems involved, the models investigated, or the interpretations of the solutions obtained or suggested. Nevertheless, researches, debates and dialogues do and should continue!

The workshop on *"Econophysics & Sociophysics of Markets & Networks"*, the third in the ECONOPHYS-KOLKATA series, was organised under the auspices of the *Centre for Applied Mathematics and Computational Science*, Saha Institute of Nuclear Physics, Kolkata, during 12–15 March, 2007, to add to the opportunity of further discussions and dialogues. As in the previous events in the series, the motto had been to have free interactions and exchanges of ideas between the various investigators: economists, financial management people, physicists, computer scientists etc. Having achieved a level of the debates and professionally accepted (though 'occasionally heated') discussions, participants clearly enjoyed the creativities involved!

This proceedings volume also indicates to the readers, who could not participate in the meeting, the level of activities and achievements by various groups. In Part I, the papers deal with the structure and also the dynamics of various financial markets. Some of the results here are reported for the first time. In Part II, the structures of various trade and business networks, including network of banks and firms are discussed in details. Again, several original observations and ideas have been reported here for the first time. In Part III, several market models are discussed. The readers will not miss the taste of the treasures of very old data and imaginative models discussed here. We also include in Part IV, in the 'Comments and Discussion' section of the book, a few summaries of the 'electrifying' discussions, criticisms and deliberations taking place during the workshop. Like in the previous volumes, this section gives a running commentary on the major issues going on in the topic.

We are grateful to all the participants of the workshop and for all their contributions. We are also extremely thankful to Mauro Gallegati and Massimo Salzano of the editorial board of **New Economic Windows** series for their support and encouragement in getting the proceedings published again in their esteemed series from Springer, Milan [Previous volumes: *Econophysics of Wealth Distributions*, Proc. Econophys-Kolkata I, Springer, Milan (2005); *Econophysics of Stock and other Markets*, Proc. Econophys-Kolkata II, Springer, Milan (2006)]. Special thanks are due to Mauro for his advisory support and suggestions on every aspect of the workshop and also to Marina Forlizzi (Springer, Milan) for her ever-ready support with the publications.

Kolkata,   *Arnab Chatterjee*
May 2007   *Bikas K. Chakrabarti*

Contents

Part III Income, Stock and Other Market Models

Part IV Comments and Discussions

List of Invited Speakers and Contributors

Dilip P. Ahalpara
Institute for Plasma Research
Near Indira Bridge
Gandhinagar-382428, India
dilip@ipr.res.in

John Angle
Inequality Process Institute
Post Office Box 429, Cabin John
Maryland, 20818, USA
angle@inequalityprocess.org

K. Bhattacharya
Satyendra Nath Bose National
Centre for Basic Sciences
Block-JD, Sector-III, Salt Lake
Kolkata-700098, India
kunal@bose.res.in

Giulio Bottazzi
Scuola Superiore Sant'Anna
Piazza Martiri della Libertà, 33
56127 Pisa, Italy
giulio.bottazzi@sssup.it

Bikas K. Chakrabarti
Theoretical Condensed Matter
Physics Division and
Centre for Applied Mathematics
and Computational Science

Saha Institute of Nuclear Physics
1/AF Bidhannagar
Kolkata 700064, India
bikask.chakrabarti@saha.ac.in

A. Chakraborti
Department of Physics
Banaras Hindu University
Varanasi-221 005, India
achakraborti@yahoo.com

A.M. Chmiel
Faculty of Physics and
Center of Excellence
for Complex Systems Research
Warsaw University of Technology
Koszykowa 75, PL 00-662 Warsaw
Poland

Giulia De Masi
Dipartimento di Economia
Università Politecnica delle Marche
P.le Martelli 8, 60121 Ancona, Italy
g.demasi@univpm.it

Zengru Di
Department of Systems Science
School of Management
Beijing Normal University
Beijing 100875, P.R. China
zdi@bnu.edu.cn

Ying Fan
Department of Systems Science
School of Management
Beijing Normal University
Beijing 100875, P.R. China
yfan@bnu.edu.cn

Yoshi Fujiwara
NiCT/ATR CIS
Applied Network Science Lab
Kyoto 619-0288, Japan
yfujiwar@atr.jp

Mauro Gallegati
Dipartimento di Economia
Università Politecnica delle Marche
P.le Martelli 8, 60121 Ancona, Italy
m.gallegati@univpm.it

Nachi Gupta
Oxford University
Computing Laboratory
Numerical Analysis Group
Wolfson Building, Parks Road
Oxford OX1 3QD, U.K.
nachi@comlab.ox.ac.uk

Raphael Hauser
Oxford University
Computing Laboratory
Numerical Analysis Group
Wolfson Building, Parks Road
Oxford OX1 3QD, U.K.

J.A. Hołyst
Faculty of Physics and
Center of Excellence
for Complex Systems Research
Warsaw University of Technology
Koszykowa 75, PL 00-662 Warsaw
Poland
jholyst@if.pw.edu.pl

Neil F. Johnson
Oxford University
Department of Physics
Clarendon Building
Parks Road, Oxford OX1 3PU, U.K.
n.johnson@physics.ox.ac.uk

Taisei Kaizoji
Division of Social Sciences
International Christian University
Mitaka, Tokyo 181-8585 Japan
kaizoji@icu.ac.jp

Menghui Li
Department of Systems Science
School of Management
Beijing Normal University
Beijing 100875, P.R. China
limh@mail.bnu.edu.cn

S.S. Manna
Satyendra Nath Bose National
Centre for Basic Sciences
Block-JD, Sector-III, Salt Lake
Kolkata-700098, India
manna@bose.res.in

K.B.K. Mayya
Physical Research Laboratory
Navrangpura, Ahmedabad-380009
India

Jürgen Mimkes
Department Physik
Universität Paderborn
Warburgerstr. 100, Germany
Juergen.Mimkes@uni-paderborn.de

Manipushpak Mitra
Economic Research Unit
Indian Statistical Institute
Kolkata
mmitra@isical.ac.in

G. Mukherjee
Satyendra Nath Bose National
Centre for Basic Sciences
Block-JD, Sector-III, Salt Lake
Kolkata-700098, India
  and
Bidhan Chandra College
Asansol 713304, Dt. Burdwan
West Bengal, India
gautamm@bose.res.in

Raj Kumar Pan
The Institute
of Mathematical Sciences
C.I.T. Campus, Taramani
Chennai – 600 113, India
rajkp@imsc.res.in

Prasanta K. Panigrahi
Physical Research Laboratory
Navrangpura, Ahmedabad-380009
India
prasanta@prl.res.in

Jitendra C. Parikh
Physical Research Laboratory
Navrangpura, Ahmedabad-380009
India
parikh@prl.res.in

M. Patriarca
Institute of Theoretical Physics
Tartu University, Tähe 4
51010 Tartu, Estonia
marco.patriarca@mac.com

Przemysław Repetowicz
Probability Dynamics
IFSC House, Custom House Quay
Dublin 1, Ireland

Peter Richmond
Department of Physics
Trinity College Dublin 2, Ireland
richmond@tcd.ie

M.S. Santhanam
Physical Research Laboratory
Navrangpura, Ahmedabad-380009
India
santh@prl.res.in

Sandro Sapio
Scuola Superiore Sant'Anna
Piazza Martiri della Libertà, 33
56127 Pisa, Italy
  and
Università di Napoli "Parthenope"
Via Medina, 40, 80133 Napoli, Italy
alessandro.sapio
  @uniparthenope.it

Abhirup Sarkar
Economic Research Unit
Indian Statistical Institute, Kolkata
abhirup@isical.ac.in

Aki-Hiro Sato
Department of Applied
Mathematics and Physics
Graduate School of Informatics
Kyoto University
Kyoto 606-8501, Japan
aki@i.kyoto-u.ac.jp

Kohei Shintani
Department of Applied
Mathematics and Physics
Graduate School of Informatics
Kyoto University
Kyoto 606-8501, Japan

J. Sienkiewicz
Faculty of Physics and
Center of Excellence
for Complex Systems Research
Warsaw University of Technology
Koszykowa 75, PL 00-662 Warsaw
Poland

Sitabhra Sinha
The Institute
of Mathematical Sciences
C.I.T. Campus, Taramani
Chennai – 600 113, India
sitabhra@imsc.res.in

Wataru Souma
NiCT/ATR CIS
Applied Network Science Lab.
Kyoto, 619-0288, Japan
souma@nict.go.jp,souma@atr.jp

Nisheeth Srivastava
The Institute
of Mathematical Sciences
C.I.T. Campus, Taramani
Chennai – 600 113, India.

K. Suchecki
Faculty of Physics and
Center of Excellence

for Complex Systems Research
Warsaw University of Technology
Koszykowa 75, PL 00-662 Warsaw
Poland

Yougui Wang
Center for Polymer Studies
Department of Physics
Boston University
Boston, MA 02215, USA
    and
Department of Systems Science
School of Management
Beijing Normal University
Beijing, 100875, P.R. China
ygwang@bnu.edu.cn

Jinshan Wu
Department of Physics & Astronomy
University of British Columbia
Vancouver, B.C. Canada, V6T 1Z1
jinshanw@phas.ubc.ca

ECONOPHYS-KOLKATA III
Econophysics & Sociophysics of Markets & Networks
12-15 March 2007

Part I

Financial Markets

Uncovering the Internal Structure of the Indian Financial Market: Large Cross-correlation Behavior in the NSE

Sitabhra Sinha and Raj Kumar Pan

The Institute of Mathematical Sciences, C.I.T. Campus, Taramani,
Chennai – 600 113, India
sitabhra@imsc.res.in

The cross-correlations between price fluctuations of 201 frequently traded stocks in the National Stock Exchange (NSE) of India are analyzed in this paper. We use daily closing prices for the period 1996–2006, which coincides with the period of rapid transformation of the market following liberalization. The eigenvalue distribution of the cross-correlation matrix, $\mathbf{C}$, of NSE is found to be similar to that of developed markets, such as the New York Stock Exchange (NYSE): the majority of eigenvalues fall within the bounds expected for a random matrix constructed from mutually uncorrelated time series. Of the few largest eigenvalues that deviate from the bulk, the largest is identified with market-wide movements. The intermediate eigenvalues that occur between the largest and the bulk have been associated in NYSE with specific business sectors with strong intra-group interactions. However, in the Indian market, these deviating eigenvalues are comparatively very few and lie much closer to the bulk. We propose that this is because of the relative lack of distinct sector identity in the market, with the movement of stocks dominantly influenced by the overall market trend. This is shown by explicit construction of the interaction network in the market, first by generating the minimum spanning tree from the unfiltered correlation matrix, and later, using an improved method of generating the graph after filtering out the market mode and random effects from the data. Both methods show, compared to developed markets, the relative absence of clusters of co-moving stocks that belong to the same business sector. This is consistent with the general belief that emerging markets tend to be more correlated than developed markets.

1 Introduction

"Because nothing is completely certain but subject to fluctuations, it
is dangerous for people to allocate their capital to a single or a small

number of securities. [...] No one has reason to expect that all securities ... will cease to pay off at the same time, and the entire capital
be lost." – *from the 1776 prospectus of an early mutual fund in the
Netherlands* [1]

As evident from the above quotation, the correlation between price movements
of different stocks has long been a topic of vital interest to those involved with
the study of financial markets. With the recent understanding of such markets
as examples of complex systems with many interacting components, these
cross-correlations have been used to infer the existence of collective modes in
the underlying dynamics of stock prices. It is natural to expect that stocks
which strongly interact with each other will have correlated price movements.
Such interactions may arise because the companies belong to the same business
sector (i.e., they compete for the same set of customers and face similar market
conditions), or they may belong to related sectors (e.g., automobile and energy
sector stocks would be affected similarly by rise in gasoline prices), or they
may be owned by the same business house and therefore perceived by investors
to be linked. In addition, all stocks may respond similarly to news breaks that
affect the entire market (e.g., the outbreak of a war) and this induces market-
wide correlations. On the other hand, information that is related only to
a particular company will tend to decorrelate its price movement from those
of others.

Thus, the effects governing the cross-correlation behavior of stock price
fluctuations can be classified into (i) market (i.e., common to all stocks), (ii)
sector (i.e., related to a particular business sector) and (iii) idiosyncratic (i.e.,
limited to an individual stock). The empirically obtained correlation structure
can then be analyzed to find out the relative importance of such effects in actual markets. Physicists investigating financial market structure have focussed
on the spectral properties of the correlation matrix, with pioneering studies
investigating the deviation of these properties from those of a random matrix,
which would have been obtained had the price movements been uncorrelated.
It was found that the bulk of the empirical eigenvalue distribution matches
fairly well with those expected from a random matrix, as does the distribution
of eigenvalue spacings [2, 3]. Among the few large eigenvalues that deviated
from the random matrix predictions, the largest represent the influence of the
entire market common to all stocks, while the remaining eigenvalues correspond to different business sectors [4], as indicated by the composition of the
corresponding eigenvectors [5]. However, although models in which the market is assumed to be composed of several correlated groups of stocks is found
to reproduce many spectral features of the empirical correlation matrix [6],
one needs to filter out the effects of the market-wide signal as well as noise
in order to identify the group structure in an actual market. Recently, such
filtered matrices have been used to reveal significant clustering among a large
number of stocks from the NYSE [7].

The discovery of complex market structure in developed financial markets as NYSE and Japan [8], brings us to the question of whether emerging markets show similar behavior. While it is generally believed that stock prices in developing markets tend to be relatively more correlated than the developed ones [9], there have been very few studies of the former in terms of analysing the spectral properties of correlation matrices [10–14][1].

In this paper we present the first detailed study of cross-correlations in the Indian financial market over a significant period of time, that coincides with the decade of rapid transformation of the recently liberalized economy into one of the fastest growing in the world. The prime motivation for our study of one of the largest emerging markets is to see if there are significant deviations from developed markets in terms of the properties of its collective modes. As already shown by us [14–16] the return distribution in Indian markets follows closely the "inverse cubic law" that has been reported in developed markets. If therefore, deviations are observed in the correlation properties, these would be almost entirely due to differences in the nature of interactions between stocks. Indeed, we do observe that the Indian market shows a higher degree of correlation compared to, e.g., NYSE. We present the hypothesis that this is due to the dominance of the market-wide signal and relative absence of significant group structure among the stocks. This may indicate that one of the hallmarks of the transition of a market from emerging to developed status is the appearance and consolidation of distinct business sector identities.

2 The Indian Financial Market

There are 23 different stock markets in India. The largest of these is the National Stock Exchange (NSE) which accounted for more than half of the entire combined turnover for all Indian financial markets in 2003–04 [17], although its market capitalization is comparable to that of the second largest market, the Bombay Stock Exchange. The NSE is considerably younger than most other Indian markets, having commenced operations in the capital (equities) market from Nov 1994. However, as of 2004, it is already the world's third largest stock exchange (after NASDAQ and NYSE) in terms of transactions [17]. It is thus an excellent source of data for studying the correlation structure of price movements in an emerging market.

Description of the data set. We have considered the daily closing price time series of stocks traded in the NSE available from the exchange web-site [18]. For cross-correlation analysis, we have focused on daily closing price data of $N = 201$ NSE stocks from Jan 1, 1996 to May 31, 2006, which corresponds to

---

[1] Most studies of correlated price movements in emerging markets have looked at *synchronicity* which measures the incidence of similar (i.e., up or down) price movements across stocks, and is not the same as correlation which measures relative magnitude of the change as well as its direction, although the two are obviously closely related.

$T = 2607$ working days (the individual stocks, along with the business sector to which they belong, are given in Table 1). The selection of the stocks was guided by the need to minimise missing data in the time-series, a problem common to data from other emerging markets [10]. In our data, 45 stocks have no missing data, while from the remaining stocks, the one having the largest fraction of missing data has price data missing for less than 6% of the total period covered[2].

3 The Return Cross-Correlation Matrix

To measure correlation between the price movements across different stocks, we first need to measure the price fluctuations such that the result is independent of the scale of measurement. For this, we calculate the logarithmic return of price. If $P_i(t)$ is the stock price of the ith stock at time t, then the (logarithmic) price return is defined as

$$R_i(t, \Delta t) \equiv \ln P_i(t + \Delta t) - \ln P_i(t). \tag{1}$$

For daily return, $\Delta t = 1$ day. By subtracting the average return and dividing the result with the standard deviation of the returns (which is a measure of the volatility of the stock), $\sigma_i = \sqrt{\langle R_i^2 \rangle - \langle R_i \rangle^2}$, we obtain the normalized price return,

$$r_i(t, \Delta t) \equiv \frac{R_i - \langle R_i \rangle}{\sigma_i}, \tag{2}$$

where $\langle \ldots \rangle$ represents time average. Once the return time series for N stocks over a period of T days are obtained, the cross-correlation matrix $\mathbf{C}$ is calculated, whose element $C_{ij} = \langle r_i r_j \rangle$, represents the correlation between returns for stocks i and j.

   If the time series are uncorrelated, then the resulting random correlation matrix, also known as a Wishart matrix, has eigenvalues distributed according to [19]:

$$P(\lambda) = \frac{Q}{2\pi} \frac{\sqrt{(\lambda_{max} - \lambda)(\lambda - \lambda_{min})}}{\lambda}, \tag{3}$$

with $N \to \infty, T \to \infty$ such that $Q = T/N \geq 1$. The bounds of the distribution are given by $\lambda_{max} = [1 + (1/\sqrt{Q})]^2$ and $\lambda_{min} = [1 - (1/\sqrt{Q})]^2$. For the NSE data, $Q = 12.97$, which implies that the distribution should be bounded at $\lambda_{max} = 1.63$ in the absence of any correlations. As seen in Fig. 1 (left), the bulk of the empirical eigenvalue distribution indeed occurs below this value. However, a small fraction ($\simeq 3$ %) of the eigenvalues deviate from the random matrix behavior, and, by analyzing them we should be able to obtain an understanding of the interaction structure of the market.

---

[2] In case of a date with missing price data, it is assumed that no trading took place on that day, so that, the price remained the same as the preceding day.

Table 1. The list of 201 stocks in NSE analyzed in this paper.

i	Company	Sector	i	Company	Sector
1	UCALFUEL	Automobiles Transport	61	SUPPETRO	Energy
2	MICO	Automobiles Transport	62	DCW	Energy
3	SHANTIGEAR	Automobiles Transport	63	CHEMPLAST	Energy
4	LUMAXIND	Automobiles Transport	64	RELIANCE	Energy
5	BAJAJAUTO	Automobiles Transport	65	HINDPETRO	Energy
6	HEROHONDA	Automobiles Transport	66	BONGAIREFN	Energy
7	MAHSCOOTER	Automobiles Transport	67	BPCL	Energy
8	ESCORTS	Automobiles Transport	68	IBP	Energy
9	ASHOKLEY	Automobiles Transport	69	ESSAROIL	Energy
10	M&M	Automobiles Transport	70	VESUVIUS	Energy
11	EICHERMOT	Automobiles Transport	71	NOCIL	Basic Materials
12	HINDMOTOR	Automobiles Transport	72	GOODLASNER	Basic Materials
13	PUNJABTRAC	Automobiles Transport	73	SPIC	Basic Materials
14	SWARAJMAZD	Automobiles Transport	74	TIRUMALCHM	Basic Materials
15	SWARAJENG	Automobiles Transport	75	TATACHEM	Basic Materials
16	LML	Automobiles Transport	76	GHCL	Basic Materials
17	VARUNSHIP	Automobiles Transport	77	GUJALKALI	Basic Materials
18	APOLLOTYRE	Automobiles Transport	78	PIDILITIND	Basic Materials
19	CEAT	Automobiles Transport	79	FOSECOIND	Basic Materials
20	GOETZEIND	Automobiles Transport	80	BASF	Basic Materials
21	MRF	Automobiles Transport	81	NIPPONDENR	Basic Materials
22	IDBI	Financial	82	LLOYDSTEEL	Basic Materials
23	HDFCBANK	Financial	83	HINDALC0	Basic Materials
24	SBIN	Financial	84	SAIL	Basic Materials
25	ORIENTBANK	Financial	85	TATAMETALI	Basic Materials
26	KARURVYSYA	Financial	86	MAHSEAMLES	Basic Materials
27	LAKSHVILAS	Financial	87	SURYAROSNI	Basic Materials
28	IFCI	Financial	88	BILT	Basic Materials
29	BANKRAJAS	Financial	89	TNPL	Basic Materials
30	RELCAPITAL	Financial	90	ITC	Consumer Goods
31	CHOLAINV	Financial	91	VSTIND	Consumer Goods
32	FIRSTLEASE	Financial	92	GODFRYPHLP	Consumer Goods
33	BAJAUTOFIN	Financial	93	TATATEA	Consumer Goods
34	SUNDARMFIN	Financial	94	HARRMALAYA	Consumer Goods
35	HDFC	Financial	95	BALRAMCHIN	Consumer Goods
36	LICHSGFIN	Financial	96	RAJSREESUG	Consumer Goods
37	CANFINHOME	Financial	97	KAKATCEM	Consumer Goods
38	GICHSGFIN	Financial	98	SAKHTISUG	Consumer Goods
39	TFCILTD	Financial	99	DHAMPURSUG	Consumer Goods
40	TATAELXSI	Technology	100	BRITANNIA	Consumer Goods
41	MOSERBAER	Technology	101	SATNAMOVER	Consumer Goods
42	SATYAMCOMP	Technology	102	INDSHAVING	Consumer Goods
43	ROLTA	Technology	103	MIRCELECTR	Consumer Discretonary
44	INFOSYSTCH	Technology	104	SURAJDIAMN	Consumer Discretonary
45	MASTEK	Technology	105	SAMTEL	Consumer Discretonary
46	WIPRO	Technology	106	VDOCONAPPL	Consumer Discretonary
47	BEML	Technology	107	VDOCONINTL	Consumer Discretonary
48	ALFALAVAL	Technology	108	INGERRAND	Consumer Discretonary
49	RIIL	Technology	109	ELGIEQUIP	Consumer Discretonary
50	GIPCL	Energy	110	KSBPUMPS	Consumer Discretonary
51	CESC	Energy	111	NIRMA	Consumer Discretonary
52	TATAPOWER	Energy	112	VOLTAS	Consumer Discretonary
53	GUJRATGAS	Energy	113	KECINTL	Consumer Discretonary
54	GUJFLUORO	Energy	114	TUBEINVEST	Consumer Discretonary
55	HINDOILEXP	Energy	115	TITAN	Consumer Discretonary
56	ONGC	Energy	116	ABB	Industrial
57	COCHINREFN	Energy	117	BHEL	Industrial
58	IPCL	Energy	118	THERMAX	Industrial
59	FINPIPE	Energy	119	SIEMENS	Industrial
60	TNPETRO	Energy	120	CROMPGREAV	Industrial

Table 1. (continued)

i	Company	Sector	i	Company	Sector
121	HEG	Industrial	161	HIMACHLFUT	Telecom
122	ESABINDIA	Industrial	162	MTNL	Telecom
123	BATAINDIA	Industrial	163	BIRLAERIC	Telecom
124	ASIANPAINT	Industrial	164	INDHOTEL	Services
125	ICI	Industrial	165	EIHOTEL	Services
126	BERGEPAINT	Industrial	166	ASIANHOTEL	Services
127	GNFC	Industrial	167	HOTELEELA	Services
128	NAGARFERT	Industrial	168	FLEX	Services
129	DEEPAKFERT	Industrial	169	ESSELPACK	Services
130	GSFC	Industrial	170	MAX	Services
131	ZUARIAGRO	Industrial	171	COSMOFILMS	Services
132	GODAVRFERT	Industrial	172	DABUR	Health Care
133	ARVINDMILL	Industrial	173	COLGATE	Health Care
134	RAYMOND	Industrial	174	GLAXO	Health Care
135	HIMATSEIDE	Industrial	175	DRREDDY	Health Care
136	BOMDYEING	Industrial	176	CIPLA	Health Care
137	NAHAREXP	Industrial	177	RANBAXY	Health Care
138	MAHAVIRSPG	Industrial	178	SUNPHARMA	Health Care
139	MARALOVER	Industrial	179	IPCALAB	Health Care
140	GARDENSILK	Industrial	180	PFIZER	Health Care
141	NAHARSPG	Industrial	181	EMERCK	Health Care
142	SRF	Industrial	182	NICOLASPIR	Health Care
143	CENTENKA	Industrial	183	SHASUNCHEM	Health Care
144	GUJAMBCEM	Industrial	184	AUROPHARMA	Health Care
145	GRASIM	Industrial	185	NATCOPHARM	Health Care
146	ACC	Industrial	186	HINDLEVER	Miscellaneous
147	INDIACEM	Industrial	187	CENTURYTEX	Miscellaneous
148	MADRASCEM	Industrial	188	EIDPARRY	Miscellaneous
149	UNITECH	Industrial	189	KESORAMIND	Miscellaneous
150	HINDSANIT	Industrial	190	ADANIEXPO	Miscellaneous
151	MYSORECEM	Industrial	191	ZEETELE	Miscellaneous
152	HINDCONS	Industrial	192	FINCABLES	Miscellaneous
153	CARBORUNIV	Industrial	193	RAMANEWSPR	Miscellaneous
154	SUPREMEIND	Industrial	194	APOLLOHOSP	Miscellaneous
155	RUCHISOYA	Industrial	195	THOMASCOOK	Miscellaneous
156	BHARATFORG	Industrial	196	POLYPLEX	Miscellaneous
157	GESHIPPING	Industrial	197	BLUEDART	Miscellaneous
158	SUNDRMFAST	Industrial	198	GTCIND	Miscellaneous
159	SHYAMTELE	Telecom	199	TATAVASHIS	Miscellaneous
160	ITI	Telecom	200	CRISIL	Miscellaneous
			201	INDRAYON	Miscellaneous

The random nature of the smaller eigenvalues is also indicated by an observation of the distribution of the corresponding eigenvector components. Note that, these components are normalized for each eigenvalue λ_j such that, $\sum_{i=1}^{N}[u_{ji}]^2 = N$, where u_{ji} is the i-th component of the jth eigenvector. For random matrices generated from uncorrelated time series, the distribution of the eigenvector components is given by the Porter-Thomas distribution,

$$P(u) = \frac{1}{\sqrt{2\pi}} \exp -\frac{u^2}{2}. \tag{4}$$

As shown in Fig. 1 (right), this distribution fits the empirical histogram of the eigenvector components for the eigenvalues belonging to the bulk. However,

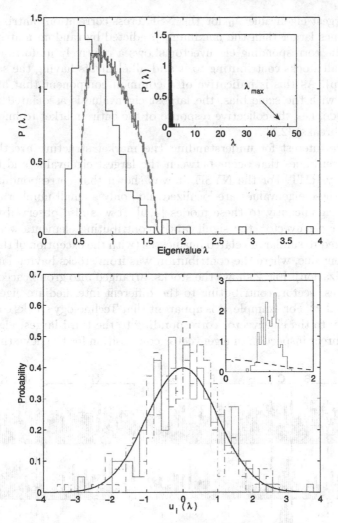

Fig. 1. (*left*) The probability density function of the eigenvalues of the cross-correlation matrix **C** for 201 stocks in the NSE of India for the period Jan 1996–May 2006. For comparison the theoretical distribution predicted by Eq. (3) is shown using broken curves, which overlaps with the spectral distribution of the surrogate correlation matrix generated by randomly shuffling the time series. The inset shows the largest eigenvalue corresponding to the market. (*Right*) The distribution of eigenvector components corresponding to three eigenvalues belonging to the bulk predicted by RMT and (inset) corresponding to the largest eigenvalue. In both cases, the Gaussian distribution expected from RMT is shown for comparison.

the eigenvectors of the largest eigenvalues (e.g., the largest eigenvalue λ_{max}, as shown in the inset) deviate quite significantly, indicating their non-random nature.

The largest eigenvalue λ_0 for the NSE cross-correlation matrix is more than 28 times larger than the maximum predicted by random matrix theory (RMT). The corresponding eigenvector shows a relatively uniform composition, with all stocks contributing to it and all elements having the same sign (Fig. 2, top). As this is indicative of a common component that affects all the stocks with the same bias, the largest eigenvalue is associated with the market mode, i.e., the collective response of the entire market to information (e.g., newsbreaks) [2,3].

Of more interest for understanding the market structure are the intermediate eigenvalues that occur between the largest eigenvalue and the bulk predicted by RMT. For the NYSE, it was shown that corresponding eigenvectors of these eigenvalues are localized, i.e., only a small number of stocks contribute significantly to these modes [4,5]. It was also observed that, for a particular eigenvector, the significantly contributing elements were stocks that belonged to similar or related businesses (with the exception of the second largest eigenvalue, where the contribution was from stocks having large market capitalization). Fig. 2 shows the stocks, arranged into groups according to their business sector, contributing to the different intermediate eigenvectors very unequally[3]. For example, it is apparent that Technology stocks contribute significantly to the eigenvector corresponding to the third largest eigenvalue. However, direct inspection of eigenvector composition for the deviating eigen-

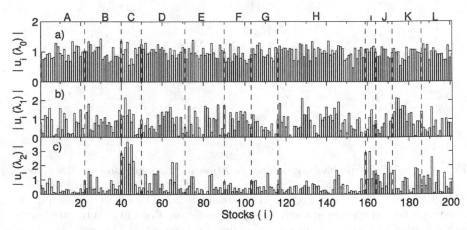

Fig. 2. The absolute values of the eigenvector components $u_i(\lambda)$ for the three largest eigenvalues of the correlation matrix $\mathbf{C}$. The stocks i are arranged by business sectors separated by *broken lines*. A: Automobile & transport, B: Financial, C: Technology, D: Energy, E: Basic materials, F: Consumer goods, G: Consumer discretionary, H: Industrial, I: IT & Telecom, J: Services, K: Healthcare & Pharmaceutical, L: Miscellaneous.

---

[3] The significant contributions to the second largest eigenvalue were found to be from the stocks SBIN, SATYAMCOMP, SURYAROSNI, ITC, BHEL, NAGARFERT, ACC, GLAXO, DRREDDY and RANBAXY.

values does not yield a straightforward interpretation of the significant group
of stocks, possibly because the largest eigenmode corresponding to the market
dominates over all intra-group correlations.

For more detailed analysis of the eigenvector composition, we use the inverse participation ratio (IPR), which is defined for the j-th eigenvector as
$I_j = \sum_{i=1}^{N} [u_{ji}]^4$, where u_{ji} are the component of jth eigenvector. For an
eigenvector with equal components, $u_{ji} = 1/\sqrt{N}$, which is approximately the
case for the eigenvector corresponding to the largest eigenvalue, $I_j = 1/N$.
If, on the other hand, a single component has a dominant contribution, e.g.,
$u_{j1} = 1$ and $u_{ji} = 0$ for $i \neq 1$, we have $I_j = 1$. Therefore, IPR is inversely
related to the number of significantly contributing eigenvector components.
For the eigenvectors corresponding to eigenvalues of a random correlation matrix, $\langle I \rangle \simeq 3/N$. As seen from Fig. 3, the eigenvalues belonging to the bulk
predicted by random matrix theory indeed have eigenvectors with this value
of IPR. But, at the lower and higher end of eigenvalues, the market shows
deviations from this value, suggesting the existence of localized eigenvectors[4].
These deviations are, however, much less significant and far fewer in number
in the Indian market compared to developed markets, implying that while
correlated groups of stocks do exist in the latter, their existence is far less
clear in the NSE.

In order to graphically present the interaction structure of the stocks in
NSE, we use a method suggested by Mantegna [20] to transform the correlation
between stocks into distances to produce a connected network in which co-

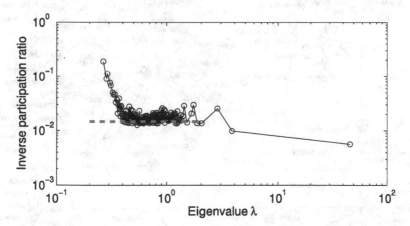

Fig. 3. Inverse participation ratio (IPR) for the different eigenvalues of the NSE
cross-correlation matrix. The *broken line* showing IPR = $3/N$ ($N = 201$, is the
number of stocks) is the expected value for a random matrix constructed from N
mutually uncorrelated time series.

---

[4] The deviations for the smallest eigenvalues indicate strong correlations between
a few stocks (see Table 2).

moving stocks are clustered together. The distance d_{ij} between two stocks i and j are calculated from the cross-correlation matrix $\mathbf{C}$, according to $d_{ij} = \sqrt{2(1 - C_{ij})}$. These are used to construct a minimum spanning tree, which connects all the N nodes of a network with $N - 1$ edges such that the total

Table 2. Stocks with dominant contribution to the six smallest eigenvalues.

λ_{201}	λ_{200}	λ_{199}	λ_{198}	λ_{197}	λ_{196}
SBIN	SBIN	RELCAPITAL	RELCAPITAL	HINDPETRO	HINDPETRO
TATAELXSI	ORIENTBANK	VDOCONAPPL	BPCL	BPCL	BPCL
ROLTA	TATAELXSI	VDOCONINTL	VDOCONAPPL	VDOCONINTL	GNFC
	ROLTA		VDOCONINTL	GNFC	GSFC
	ACC		NAHARSPG	NAHARSPG	NAHAREXP
					NAHARSPG
					ESSELPACK

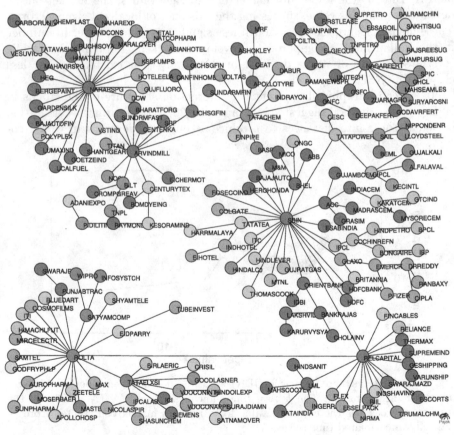

Fig. 4. The minimum spanning tree connecting 201 stocks of NSE. The *node colors* indicate the business sector to which a stock belongs. The figure has been drawn using the Pajek software. Contact authors for color figures.

sum of the distance between every pair of nodes, $\sum_{i,j} d_{ij}$, is minimum. For the NYSE, such a construction has been shown to cluster together stocks belonging to the same business sector [21]. However, as seen in Fig. 4, for the NSE, such a method fails to clearly segregate any of the business sectors. Instead, stocks belonging to very different sectors are equally likely to be found within each cluster. This suggests that the market mode is dominating over all intra-sector interactions.

Therefore, to be able to identify the internal structure of interactions between the stocks we need to remove the market mode, i.e., the effect of the largest eigenvalue. Also, the effect of random noise has to be filtered out. To perform this filtering, we use the method proposed in Ref. [7] where the correlation matrix was expanded in terms of its eigenvalues λ_i and the corresponding eigenvectors $\mathbf{u}_i$: $\mathbf{C} = \Sigma_i \lambda_i \mathbf{u}_i \mathbf{u}_i^T$. This allows the correlation matrix to be decomposed into three parts, corresponding to the market, sector and random components:

$$\mathbf{C} = \mathbf{C}_{market} + \mathbf{C}_{sector} + \mathbf{C}_{random} = \lambda_0 \mathbf{u}_0^T \mathbf{u}_0 + \sum_{i=1}^{N_s} \lambda_i \mathbf{u}_i^T \mathbf{u}_i + \sum_{i=N_s+1}^{N-1} \lambda_i \mathbf{u}_i^T \mathbf{u}_i,$$

(5)

where, the eigenvalues have been arranged in descending order (the largest labelled 0) and N_s is the number of intermediate eigenvalues. From the empirical data, it is not often obvious what is the value of N_s, as the bulk may deviate from the predictions of random matrix theory because of underlying structure induced correlations. For this reason, we use visual inspection of the distribution to choose $N_s = 5$, and verify that small changes in this value does not alter the results. The robustness of our results to small variations in the estimation of N_s is because the error involved is only due to the eigenvalues closest to the bulk that have the smallest contribution to $\mathbf{C}_{sector}$. Fig. 5 shows the result of the decomposition of the full correlation matrix into the three components. Compared to the NYSE, NSE shows a less extended tail for the sector correlation matrix elements C_{ij}^{sector}. This implies that the Indian market has a much smaller fraction of strongly interacting stocks, which would be the case if there is no significant segregation into sectors in the market.

Next, we construct the network of interactions among stocks by using the information in the sector correlation matrix [7]. The binary-valued adjacency matrix $\mathbf{A}$ of the network is generated from $\mathbf{C}_{sector}$ by using a threshold c_{th} such that $A_{ij} = 1$ if $C_{ij}^{sector} > c_{th}$, $A_{ij} = 0$ otherwise. If the long tail in the C_{ij}^{sector} distribution is indeed due to correlations among stocks belonging to a particular business sector, this should be reflected in a clustered structure of the network for an appropriate choice of the threshold. Fig. 6 shows the resultant network for the best choice of $c_{th} = c^*$ ($= 0.09$) in terms of creating the largest clusters of related stocks. However, even for the "best" choice we find that only two sectors have been properly clustered, those corresponding to Technology and to Pharmaceutical Companies. The majority of the frequently traded stocks cannot be arranged into well-segregated groups corresponding

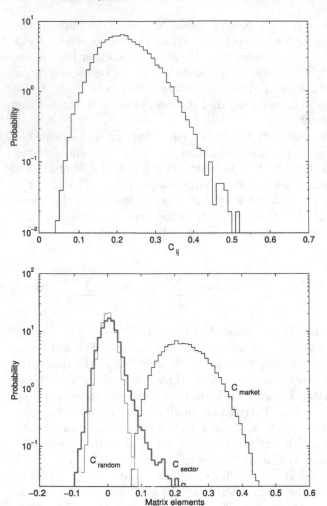

Fig. 5. (*left*) The distribution for the components C_{ij} of the cross-correlation matrix for NSE. (*Right*) The matrix element distributions following decomposition of $\mathbf{C}$ into sector, $\mathbf{C}_{sector}$, market, $\mathbf{C}_{market}$, and random effects, $\mathbf{C}_{random}$, with $N_s = 5$.

to the various business sectors they belong to. This failure again reflects the fact that intra-group correlations in most cases are much weaker compared to the market-wide correlation in the Indian market.

4 Time-evolution of the Correlation Structure

In this section, we study the temporal properties of the correlation matrix. We note here that if the deviations from the random matrix predictions are indicators of genuine correlations, then the eigenvectors corresponding to the deviating eigenvalues should be stable in time, over the period used to calculate the correlation matrix. We choose the eigenvectors corresponding to the

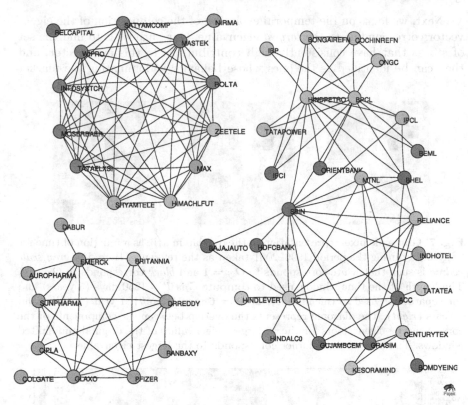

Fig. 6. The network of stock interactions in NSE generated from the group correlation matrix $\mathbf{C}_{sector}$ with threshold $c^* = 0.09$. The *node colors* indicate the business sector to which a stock belongs. The *top left cluster* comprises mostly Technology stocks, while the *bottom left cluster* is composed almost entirely of Healthcare & Pharmaceutical stocks. By contrast, the *larger cluster on the right* is not dominated by any particular sector. The figure has been drawn using the Pajek software. Contact the authors for color figures.

10 largest eigenvalues for the correlation matrix over a period $A = [t, t + T]$ to construct a 10×201 matrix $\mathbf{D}_A$. A similar matrix $\mathbf{D}_B$ can be generated by using a different time period $B = [t + \tau, t + \tau + T]$ having the same duration but a time lag τ compared to the other. These are then used to generate the 10×10 overlap matrix $\mathbf{O}(t, \tau) = \mathbf{D}_A \mathbf{D}_B^T$. In the ideal case, when the 10 eigenvectors are absolutely stable in time, $\mathbf{O}$ would be a identity matrix. For the NSE data we have used time lags of $\tau = 6$ months, 1 year and 2 years, for a time window of 5 years and the reference period beginning in Jan 1996. As shown in Fig. 7 the eigenvectors show different degrees of stability, with the one corresponding to the largest eigenvalue being the most stable. The remaining eigenvectors show decreasing stability with an increase in the lag period.

Next, we focus on the temporal evolution of the composition of the eigenvector corresponding to the *largest* eigenvalue. Our purpose is to find the set of stocks that have consistently high contributions to this eigenvector, and they can be identified as the ones whose behavior is dominating the market

Fig. 7. Grayscale pixel representation of the overlap matrix as a function of time for daily data during the period 1996–2001 taken as the reference. Here, the *gray scale coding* is such that white corresponds to $O_{ij} = 1$ and *black* corresponds to $O_{ij} = 0$. The length of the time window used to compute C is $T = 1250$ days (5 years) and the separations used to calculate O_{ij} are $\tau = 6$ months (*left*), 1 year (*middle*) and 2 years (*right*). The diagonal represents the overlap between the components of the corresponding eigenvectors for the 10 largest eigenvalues of the original and shifted windows. The bottom right corner corresponds to the largest eigenvalue.

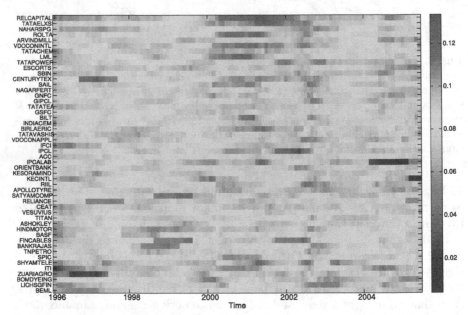

Fig. 8. The 50 stocks which have the largest contribution to the eigenvector components of the largest eigenvalue as a function of time for the period Jan 1996–May 2006. The *grayscale intensity* represents the degree of correlation. Contact the authors for color figures.

mode. We study the time-development by dividing the return time-series data into M overlapping sets of length T. Two consecutive sets are displaced relative to each other by a time lag δt. In our study, T is taken as six months (125 trading days), while δt is taken to be one month (21 trading days). The resulting correlation matrices, $\mathbf{C}_{T,\delta t}$, can now be analysed to get further understanding of the time-evolution of correlated movements among the different stocks.

In a previous paper [14], we have found that the largest eigenvalue of $\mathbf{C}_{T,\delta t}$ follows closely the time variation of the average correlation coefficient. This indicates that the largest eigenvalue λ_0 captures the behavior of the entire market. However, the relative contribution to its eigenvector $\mathbf{u}_0$ by the different stocks may change over time. We assume that if a company is a really important player in the market, then it will have a significant contribution in the composition of $\mathbf{u}_0$ over many time windows. Fig. 8 shows the 50 largest stocks in terms of consistently having large representation in $\mathbf{u}_0$. Note the existence of 5 companies from the Tata group and 3 companies of the Reliance group in this set. This is consistent with the general belief in the business community that these two groups dominate the Indian market, and may disproportionately affect the market through their actions.

5 Conclusions

In this paper, we have examined the structure of the Indian financial market through a detailed investigation of the spectral properties of the cross-correlation matrix of price returns. We demonstrate that the eigenvalue distribution is similar to that observed for developed markets of USA and Japan. However, unlike the latter, the Indian market shows much less evidence of the existence of business sectors having distinct identities. In fact, most of the observed correlation among stocks is due to effects common to the entire market, which has the effect of making the Indian market appear more correlated than developed markets. We hypothesise that the reason why emerging markets have been often reported to be significantly more correlated is because they are distinguished from developed ones in the absence of strong interactions between clusters of stocks in the former. This has implications for the understanding of markets as complex interacting systems, namely, that interactions emerge between groups of stocks as a market evolves over time to finally exhibit the clustered structure characterizing, e.g., the NYSE. How such self-organization is related to other changes a market undergoes as it develops is a question worth pursuing with the tools available to econophysicists. From the point of view of possible applicability, these results are of significance to the problem of portfolio diversification. With the advent of liberalization, there has been a significant flow of investment into the Indian market. The question of how investments can be made over a balanced portfolio of stocks so as to minimize risks assumes importance in such a situation.

Our study indicates that schemes for constructing such optimized portfolios must take into account the fact that emerging markets are in general less differentiated and more correlated than developed markets.

Acknowledgement. We thank N. Vishwanathan for assistance in preparing the data for analysis and M. Marsili for helpful discussions.

References

1. Rouwenhorst K G (2005) The origins of mutual funds. In: Goetzmann, W N, Rouwenhorst, K G (eds) The Origins of Value: The financial innovations that created modern capital markets. Oxford Univ Press, New York
2. Laloux L, Cizeau P, Bouchaud J P, Potters M (1999) Noise dressing of financial correlation matrices, Phys. Rev. Lett. 83: 1467–1470
3. Plerou V, Gopikrishnan P, Rosenow B, Amaral L A N, Stanley H E (1999) Universal and nonuniversal properties of cross correlations in financial time series, Phys. Rev. Lett. 83: 1471–1474
4. Gopikrishnan P, Rosenow B, Plerou V, Stanley H E (2001) Quantifying and interpreting collective behavior in financial markets, Phys. Rev. E 64: 035106
5. Plerou V, Gopikrishnan P, Rosenow B, Amaral L A N, Guhr T, Stanley H E (2002) Random matrix approach to cross correlations in financial data, Phys. Rev. E 65: 066126
6. Noh J D (2000) Model for correlations in stock markets, Phys. Rev. E 61: 5981–5982
7. Kim D-H, Jeong H (2005) Systematic analysis of group identification in stock markets, Phys. Rev. E 72: 046133
8. Utsugi A, Ino K, Oshikawa M (2004) Random matrix theory analysis of cross correlations in financial markets, Phys. Rev. E 70: 026110
9. Morck R, Yeung B, Yu W (2000) The information content of stock markets: Why do emerging markets have synchronous stock price movements?, J. Financial Economics 58: 215–260
10. Wilcox D, Gebbie T (2004) On the analysis of cross-correlations in South African market data, Physica A 344: 294–298; Wilcox D, Gebbie T (2007) An analysis of cross-correlations in an emerging market, Physica A 375:584–598
11. Kulkarni V, Deo N (2005) Volatility of an Indian stock market: A random matrix approach, In: Chatterjee A, Chakrabarti B K (eds) Econophysics of Stock and Other Markets. Springer, Milan, p 35
12. Jung W-S, Chaea S, Yanga J-S, Moon H-T (2006) Characteristics of the Korean stock market correlations, Physica A 361: 263–271
13. Cukur S, Eryigit M, Eryigit R (2007) Cross correlations in an emerging market financial data, Physica A 376: 555–564
14. Sinha S, Pan R K (2006) The power (law) of indian markets: Analysing NSE and BSE trading statistics, In: Chatterjee A, Chakrabarti B K (eds) Econophysics of Stock and Other Markets. Springer, Milan, pp. 24–34
15. Pan R K, Sinha S (2007) Self-organization of price fluctuation distribution in evolving markets, Europhys. Lett. 77: 58004

16. Pan R K, Sinha S (2006) Inverse cubic law of index fluctuation distribution in Indian markets, physics/0607014
17. National Stock Exchange (2004) Indian securities market: A review. (http://www.nseindia.com/content/us/ismr2005.zip)
18. http://www.nseindia.com/
19. Sengupta A M, Mitra P P (1999) Distribution of singular values for some random matrices, Phys. Rev. E 60: 3389–3392
20. Mantegna R N (1999) Hierarchical structure in financial markets, Eur. Phys. J. B 11: 193–197
21. Onnela J-P, Chakraborti A, Kaski K, Kertesz J (2002) Dynamic asset trees and portfolio analysis, Eur. Phys. J. B 30:285–288

Power Exponential Price Returns in Day-ahead Power Exchanges

Giulio Bottazzi[1] and Sandro Sapio[1,2]

[1] Scuola Superiore Sant'Anna, Piazza Martiri della Libertà, 33, 56127 Pisa, Italy
   giulio.bottazzi@sssup.it
[2] Università di Napoli "Parthenope", Via Medina, 40, 80133 Napoli, Italy
   alessandro.sapio@uniparthenope.it

Summary. This paper uses the Subbotin power exponential family of probability densities to statistically characterize the distribution of price returns in some European day-ahead electricity markets (NordPool, APX, Powernext). We implement a generic non-parametric method, known as Cholesky factor algorithm, in order to remove the strong seasonality and the linear autocorrelation structure observed in power prices. The filtered NordPool and Powernext data are characterized by an inverse relationship between the returns volatility and the price level – approximately a linear functional dependence in log-log space, which properly applied to the Cholesky residuals yields a homoskedastic sample. Finally, we use Maximum Likelihood estimation of the Subbotin family on the rescaled residuals and compare the results obtained for different markets. All empirical densities, irrespectively of the time of the day and of the market considered, are well described by a heavy-tailed member of the Subbotin family, the Laplace distribution.

Key words: Electricity Markets, Subbotin Distribution, Fat Tails, Scaling, Persistence

1 Introduction

Recent years have witnessed an improved understanding of volatility and persistence as pervasive features of the dynamics of economic activity and market exchanges. Such properties are supposed to hold *a fortiori* under circumstances such as low liquidity, low elasticity of demand, non-storability and market power. Day-ahead power pools are a case in point: they are characterized by a challenging set of empirical regularities in search of an explanation – multiple periodic patterns, persistency, spikes, heavy tails, time-dependent volatility. A cursory list of contributions within the relevant literature includes Geman and Roncoroni (2002), Eberlein and Stahl (2003), Weron, Bierbrauer and Truck (2004), Sapio (2004), Bottazzi, Sapio and Secchi (2005), Knittel and Roberts (2005), Guerci et al. (2006).

The present work provides a detailed study of some of the most interesting properties of the dynamics of prices in European day-ahead power exchanges – the Scandinavian *NordPool*, the Dutch *APX* and the French *Powernext*. The paper provides an assessment of the strength of serial correlations in the growth of prices in different hourly auctions, corresponding to different demand conditions, and a simple characterization of the behavior of volatility. Once all of these properties are controlled for, the distributional nature of the underlying statistically independent disturbances can be studied.

A more detailed account of our results is the following. *First,* we find that significant autocorrelation at weekly lags is a robust feature of power auctions when the level of economic activity (and of electricity demand) is high, whereas the degree of persistency is lower during night hours, and more heterogeneous across countries. *Second,* the conditional standard deviation of price growth is decreasing in the price level in the NordPool and Powernext markets, but this relationship breaks down when price reaches very high levels. The scaling evidence for the APX is rather mixed. *Finally,* density fit exercises, based on the Subbotin family of distributions – a family including Laplace and Normal laws as special cases – show that heavy tails are a robust feature of electricity price growth rates.

The paper is organized as follows. Section 2 offers an overview of the main empirical findings on serial correlations (2.1), volatility scaling (2.2), and distributional shapes (2.3). Remarks and conclusions are in Sect. 3.

2 Empirical Analysis of Electricity Log-returns

The following analysis aims to extend and integrate the evidence presented in [3] on serial correlation patterns, volatility structures, and distributional shapes.

Define P_{ht} as the market-clearing price issued on day t, outcome of the auction concerning delivery at hour h of the following day, and p_{ht} its natural logarithm. The daily price growth rates, or *log-returns*, are $r_{ht} = p_{ht} - p_{h,t-1} \approx P_{ht}/P_{h,t-1} - 1$.

Looking at Fig. 1 it is clear that price dynamics differ both with respect to the market (NordPool, APX or Powernext) and to the hour they refer to. All series display sharp and short-lived spikes, while the annual seasonality is more evident in NordPool prices than in APX and Powernext.

In Table 1, summary statistics for r are provided, for series regarding the 1 a.m. and the 12 a.m. auctions. Auctions for the provision of electricity by night yield similar outcomes as the 1 a.m. one, whereas the 12 a.m. well represents the properties of prices and volumes in auctions for the supply of electricity during the hours when the economic activity is higher.[1] The first three rows of Table 1 show that, while drifts and asymmetries in price

---

[1] The complete statistical information on all auctions is available upon request.

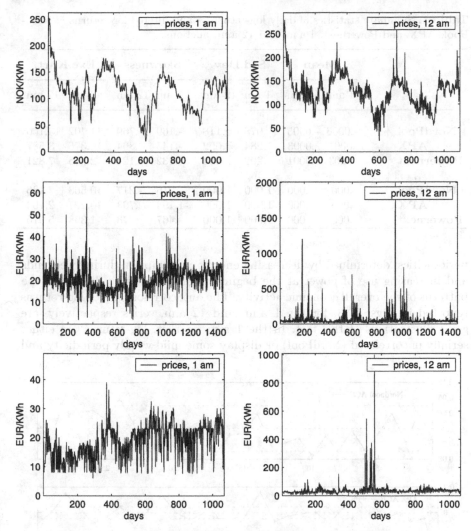

Fig. 1. *Top:* NordPool prices, from Jan 1, 1997, to Dec 31, 1999: 1 a.m. and 12 a.m. auctions. *Middle:* APX prices from Jan 6, 2001, to Dec 31, 2004: 1 a.m. and 12 a.m. auctions. *Bottom:* Powernext prices, from Feb 1, 2002, to Dec 31, 2004: 1 a.m. and 12 a.m. auctions.

growth distributions are rather weak, standard deviations are clearly higher in day-time auctions than during the night, in all countries.

2.1 The Autocorrelation Structure of Log-return

It is rather well known that fluctuations in day-ahead electricity prices are persistent and systematic (see [12], [15], and [13]). Along with yearly seasonals and long-run trends, the dynamics of prices is characterized by weekly

Table 1. Summary statistics of daily log-returns r and filtered log-returns $\tilde{r}$: Nord-Pool, APX, and Powernext, 1 a.m. and 12 a.m. auctions.

	Mean		Std.Dev.		Skewness		Exc.Kurt.	
	1 am	12 am	1 am	12 am	1 am	12 am	1 am	12 am
r								
NordPool	-.0006	-.0005	.078	.118	-.169	.769	11.402	13.100
APX	.0001	.0003	.284	.692	-.145	.804	16.376	3.587
Powernext	.0003	-.0001	.292	.494	.132	1.124	2.815	7.821
$\tilde{r}$								
NordPool	.000	.000	1.000	1.000	-.216	-.317	10.568	7.136
APX	.000	.000	1.000	1.000	-1.194	.703	10.227	2.704
Powernext	.000	.000	1.000	1.000	-.467	.273	1.398	5.976

periodicities, determined by decreasing energy consumption during weekends and increasing use of power at the beginning of the week, in line with the patterns of the overall economic activity. The autocorrelograms of log-returns reported in Figs. 2, 3, for the 1 a.m. and 12 a.m. series respectively, are good illustrations of this fact. In the 1 a.m. market, log-returns are either serially uncorrelated (NordPool) or display some mild weekly periodicity and

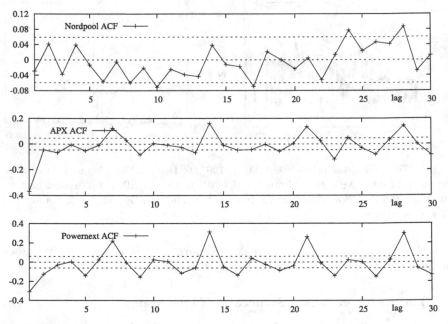

Fig. 2. Autocorrelograms of log-returns r, 1 a.m. auctions together with 95% confidence band, for NordPool (*top*), APX (*middle*) and Powernext (*bottom*).

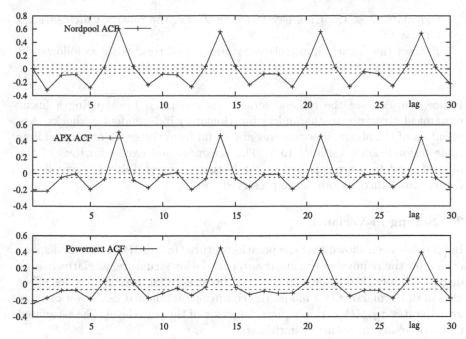

Fig. 3. Autocorrelograms of log-returns r, 1 a.m. auctions together with 95% confidence band, for NordPool (*top*), APX (*middle*) and Powernext (*bottom*).

a small, negative autocorrelation at lag 1 day (APX, Powernext). Conversely, log-returns in the 12 a.m. auctions are always strongly autocorrelated at lags 1 week, 2 weeks, and so forth. Autocorrelation coefficients are between 0.4 and 0.6 for the first weekly lags, and decay quite slowly.

The previous graphical analysis reveals that the time series of electricity prices are characterized by a rich intertemporal structure, plausibly encompassing both deterministic cyclical components and stochastic seasonal effects. Devising a linear model able to account for this intricate structure with just few parameters, albeit interesting, is in principle a very difficult task. Here we do not pursue this goal. Because we are interested in the description of the distributional properties of the underlying stochastic process, we adopt a broader non-parametric approach.

Following [3], we decompose the log-returns time series $r = \{r_t\}$ into an estimated stationary variance-covariance marix and a set of uncorrelated residuals. The filtering procedure followed here is based on the Cholesky factor algorithm, a model-free bootstrapping method described in [8]. It goes through the following steps (where indexes are omitted for clarity):

1. Estimate the covariance matrix $\Sigma = \mathrm{E}\left[r'\,r\right]$ of the vector r, as the Toeplitz matrix (that is a matrix which has constant values along all negative-sloping diagonals) built upon the autocovariance vector γ;

2. Calculate C as the Cholesky factor of Σ, i.e. the matrix which satisfies $CC' = \Sigma$;
3. Extract the linearly uncorrelated, standardized residuals $\tilde{r}$ as follows:

$$\tilde{r} = C^{-1}r \tag{1}$$

Hence, one can see the original series r as generated by applying a linear dynamical structure to the underlying (linearly) independent $\tilde{r}$ shocks. Application of the above filter removes any trend from the series and normalizes their unconditional variance to 1. The skewness and excess kurtosis of the filtered log-return are reported in the bottom rows of Table 1. Overall, the comparative kurtosis values are preserved.

2.2 Scaling in Variance

In [3] it has been shown that the price log-returns in NordPool market display, even after the removal of the linear autocorrelation structure, a relatively high degree of heteroskedasticity. In order to investigate the presence of this effect also in other markets, we model the conditional standard deviation of price growth rates $\sigma[\tilde{r}_{ht}|P_{h,t-1}]$ as a power function of the price level. The estimated equation reads (in natural logarithms)

$$\log \sigma[\tilde{r}_{ht}|P_{h,t-1}] = \chi + \rho p_{h,t-1} + \epsilon_{ht} \tag{2}$$

where χ and ρ are constant coefficients and ϵ_{ht} is an i.i.d. error term. The dependent variable is the sample standard deviation of *filtered* log-returns. Notice that the parameter ρ is null under a multiplicative random walk while it equals -1 for all additive processes (stationary or not).

Estimation of (2) is implemented by grouping, for any given time series, data into equipopulate bins. The sample standard deviations of log-returns in each bin is then computed, and its logarithm regressed (with the standard Ordinary Least Squares procedure) on a constant and on the logarithm of the mean price level within the corresponding bin. A critical issue with this estimation procedure regards the choice of the number of bins, which affects both the estimation of standard deviations in each bin (the dependent variable) and the properties of the OLS estimator of the scaling coefficients. The results of a first round of estimations suggest that the fitting performance of (2) for NordPool and Powernext can be considerably improved by including a dummy d, as follows:

$$\log \sigma[\tilde{r}_{ht}|p_{h,t-1}] = \chi + \chi' d + \rho p_{h,t-1} + \rho' p_{h,t-1} d + \epsilon_{ht} \tag{3}$$

where $d = 1$ for the last two bins (i.e. those corresponding to the highest price levels), $d = 0$ otherwise. Such a dummy allows both the slope and the intercept of the scaling regression to vary, depending on whether the price level is high or low. This amounts to assume that different scaling relationships and different volatility behaviors are at work, corresponding to diverse demand regimes.

The scatter plots of σ vs. the average (log) price level p are reported in Fig. 4, together with the fit of (3) for NordPool and Powernext, and (2) for the APX, concerning different hourly auctions.

Point estimates of the coefficient ρ are reported in Table 2 and suggest that, for NordPool and Powernext markets, standard deviations of normalized log-returns are negatively correlated with lagged price levels : $\hat{\rho} < 0$. In the NordPool, $\hat{\rho}$ fluctuates around –1, with lower absolute values in peak hours. The estimated model accounts for quite a high percentage of the dispersion in log-return volatility (see R^2 values). As to the Powernext, point estimates of the scaling exponent are rather homogeneous across auctions – approximately in the range (–0.72, –0.59) – with the exceptions of the auctions between 1 and 4 a.m. (lower absolute values), and those between 8 and 12 p.m. (higher absolute values). Finally, it is difficult to identify any clear pattern in the

Table 2. Slopes of the power law relationship between standard deviation of filtered daily log-returns and lagged price levels: NordPool (20 bins), APX (28 bins), Powernext (20 bins).

Auctions	NordPool		APX		Powernext	
	$\hat{\rho}_r$	R^2	$\hat{\rho}_r$	R^2	$\hat{\rho}_r$	R^2
1 am	$-1.087 \pm .011$	.750	$-.526 \pm .010$	.293	$-.475 \pm .008$	.563
2	$-1.193 \pm .010$	.804	$-.724 \pm .012$	.370	$-.368 \pm .008$	.495
3	$-1.022 \pm .011$	.721	$-.616 \pm .008$	.461	$-.450 \pm .007$	.598
4	$-1.088 \pm .008$	.838	$-.385 \pm .006$	.397	$-.215 \pm .008$	.307
5	$-1.143 \pm .007$	.884	$-.394 \pm .005$	.452	$-.594 \pm .011$	.437
6	$-1.207 \pm .008$	.867	$-.394 \pm .005$	.462	$-.712 \pm .007$	.730
7	$-1.172 \pm .012$	.754	$-.301 \pm .004$	.451	$-.703 \pm .006$	.830
8	$-1.076 \pm .009$	.804	$-.300 \pm .005$	.393	$-.628 \pm .006$	.723
9	$-.827 \pm .014$	.530	$-.394 \pm .006$	.384	$-.643 \pm .007$	.716
10	$-.765 \pm .014$	.532	$-.111 \pm .008$	.033	$-.797 \pm .006$	.804
11	$-.731 \pm .011$	.565	$-.021 \pm .007$	.002	$-.669 \pm .009$	.593
12	$-.914 \pm .013$	.573	$.085 \pm .004$	.054	$-.619 \pm .009$	.605
1 pm	$-1.059 \pm .010$	.743	$.142 \pm .006$	.086	$-.691 \pm .010$	.577
2	$-1.117 \pm .008$	.833	$.140 \pm .005$	.109	$-.718 \pm .008$	.685
3	$-1.096 \pm .007$	.850	$.062 \pm .006$	.018	$-.703 \pm .007$	.718
4	$-1.017 \pm .012$	.675	$.049 \pm .006$	.009	$-.661 \pm .009$	.630
5	$-.819 \pm .012$	.561	$.052 \pm .007$	.010	$-.633 \pm .007$	.716
6	$-.659 \pm .015$	.351	$.264 \pm .004$	.444	$-.570 \pm .007$	.644
7	$-.888 \pm .012$	.600	$.393 \pm .005$	.445	$-.632 \pm .010$	.495
8	$-.981 \pm .011$	.687	$.495 \pm .004$	.665	$-.831 \pm .009$	.704
9	$-1.030 \pm .010$	.727	$.298 \pm .006$	.257	$-.889 \pm .009$	.738
10	$-1.157 \pm .012$	.696	$-.585 \pm .022$	.105	$-1.049 \pm .009$	.785
11	$-1.054 \pm .012$	.675	$-.399 \pm .025$	.039	$-1.130 \pm .010$	.785
12	$-1.030 \pm .010$	.765	$-.683 \pm .020$	.154	$-1.223 \pm .011$	.789

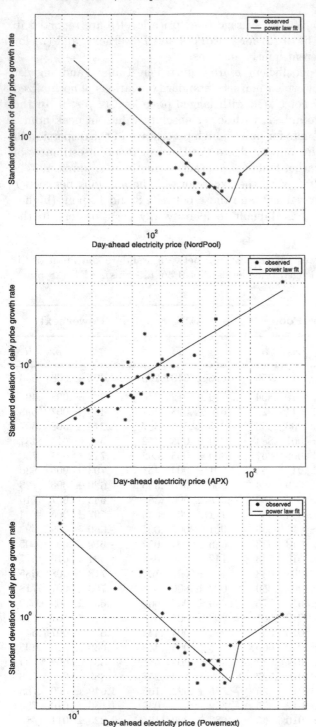

Fig. 4. Linear fit of the relationship between log of the conditional standard deviation of filtered log-returns, $\log \sigma[\widetilde{r}_t | P_{t-1}]$, and lagged log-price level, p_{t-1}. Plots are in double logarithmic scale. NordPool (*top*, 3 p.m., $\widehat{\rho}_r = -1.096$), APX (*middle*, 8 p.m., $\widehat{\rho}_r = .495$), Powernext (*bottom*, 10 a.m., $\widehat{\rho}_r = -.831$).

volatility structure of APX log-returns: $\hat{\rho} > 0$ between 12 a.m. and 6 p.m., i.e. when the economic activity is high – but the fitting performance is weak. In other hours, $\hat{\rho} < 0$ to various degrees. In sum, point estimates of the APX scaling exponents display too huge a variability, and no clear conclusion can be drawn.

In general, we can use the statistical characterization provided by (3) in order to obtain a rescaled version of $\tilde{r}$. This rescaling is necessary in order to take care of the heteroskedastic structure of filtered price returns. We define the rescaled growth rates r_{ht}^* as

$$r_{ht}^* = \frac{\tilde{r}_{ht}}{\hat{\sigma}} \tag{4}$$

where $\hat{\sigma}$ is the conditional standard deviation of log-returns, as predicted by (2) (APX) and (3) (NordPool, Powernext).

2.3 Power Exponential Distribution

As a final step of our analysis, we investigate the shape of the probability density function of price returns. We take a parametric approach and use a flexible family of probability densities known as Subbotin (or Power Exponential) family, in order to quantify the degrees of peakedness and heavy-tailedness of the empirical densities within a quite general and parsimonious framework.

First used in economics by [5], the Subbotin probability density function of a generic random variable X reads (see also [14]):

$$Pr\{X = x\} = \frac{1}{2ab^{1/b}\Gamma(1 + \frac{1}{b})}e^{-\frac{1}{b}|\frac{x-\mu}{a}|^b} \tag{5}$$

with parameters a (width), b (shape), and μ (position). $\Gamma(.)$ is the gamma function. The Subbotin reduces to a Laplace if $b = 1$ and to a Gaussian if $b = 2$. The Continuous Uniform is a limit case for $b \to \infty$. As b gets smaller, the density becomes heavier-tailed and more sharply peaked.

Compared to previously fitted distributions, such as the Generalized Hyperbolic ([9]), the Subbotin is more parsimonious: just 3 parameters need to be estimated, or 2 if the data are demeaned. It also allows for greater flexibility with respect to the tail behavior: the Generalized Hyperbolic distribution family only admits exponential tail decay (cf. the application by [9] on Nord-Pool data). On the other hand, evidence of Subbotin distributions would rule out power-law tails, which characterize Levy phenomena with tail index $\alpha < 2$ (see [2], [1] and [7], for applications to electricity price dynamics).

Estimates of the Subbotin parameters are obtained through Maximum Likelihood methods. For a general discussion of the property of the estimates see [6].[2]

---

[2] The estimates reported in this paper are obtained with the use of the Subbotools software package (see [4]).

Table 3. Estimates of the Subbotin shape parameters for daily filtered and rescaled price growth rates: NordPool, APX, and Powernext.

Auctions	NordPool r̃ $\hat{b}$	std.err.	NordPool r* $\hat{b}$	std.err.	APX r̃ $\hat{b}$	std.err.	APX r* $\hat{b}$	std.err.	Powernext r̃ $\hat{b}$	std.err.	Powernext r* $\hat{b}$	std.err.
1 am	.967	.054	1.184	.069	1.081	.054	1.023	.050	1.259	.076	1.307	.079
2	.899	.049	1.143	.066	.852	.040	.780	.036	1.226	.073	1.348	.082
3	.901	.050	1.110	.064	.865	.041	.759	.035	1.223	.073	1.341	.082
4	.911	.050	1.236	.073	.837	.039	.756	.035	1.187	.070	1.232	.074
5	.927	.051	1.311	.079	.806	.038	.720	.033	1.057	.061	1.116	.065
6	.885	.048	1.217	.072	.785	.036	.708	.032	1.085	.063	1.253	.075
7	.930	.051	1.163	.068	.754	.035	.674	.030	.948	.053	1.090	.063
8	.944	.052	1.152	.067	.778	.036	.689	.031	1.040	.060	1.188	.070
9	.956	.053	1.118	.064	.846	.040	.777	.036	.999	.057	1.161	.068
10	.995	.056	1.139	.066	.899	.043	.902	.043	.942	.053	1.096	.064
11	1.146	.066	1.311	.079	1.023	.050	1.023	.050	1.081	.063	1.194	.071
12	1.047	.059	1.204	.071	1.233	.063	1.224	.062	1.022	.059	1.114	.065
1 pm	1.070	.061	1.280	.076	1.121	.056	1.103	.055	1.049	.060	1.169	.069
2	1.089	.062	1.326	.080	1.153	.058	1.129	.057	1.073	.062	1.247	.075
3	1.059	.060	1.275	.076	1.103	.055	1.097	.055	1.042	.060	1.202	.071
4	1.072	.061	1.298	.078	1.026	.050	1.022	.050	1.060	.061	1.220	.073
5	1.048	.059	1.206	.071	1.045	.051	1.035	.051	1.220	.073	1.407	.087
6	.986	.055	1.166	.068	1.068	.053	1.000	.049	1.147	.067	1.272	.077
7	1.069	.061	1.221	.072	.976	.047	.950	.046	1.021	.058	1.124	.066
8	1.005	.057	1.175	.068	1.102	.055	1.048	.052	1.027	.059	1.096	.064
9	.936	.052	1.144	.066	1.084	.054	1.056	.052	1.148	.067	1.372	.084
10	.972	.054	1.169	.068	.811	.038	.775	.036	1.126	.066	1.328	.081
11	.998	.056	1.184	.069	.772	.036	.759	.035	1.129	.066	1.371	.084
12	1.029	.058	1.218	.072	.864	.041	.815	.038	1.063	.061	1.327	.081

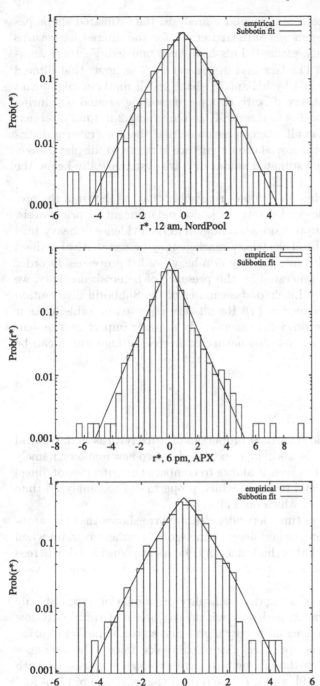

Table 3 reports, for each market under analysis, the estimated shape parameter $\hat{b}$ and the corresponding standard errors for the filtered log-returns ($\tilde{r}$). For the sake of brevity, estimated a's have been omitted. Furtherly, normalization implies $\mu = 0$. The fact that in general $\hat{b} \sim 1$ suggests that filtered log-returns are characterized by a Laplace shape in all markets. Significant deviations are however observed, with $\hat{b}$ values clustering around 0.8 during night hours and early morning in the APX, and around 1.2 in the Powernext between 1 and 3 a.m. All in all, one can conclude that the price returns distributions, once the linear autocorrelation structure is removed, display heavy-tailed shapes. Indeed, the estimated values of $\hat{b}$ are systematically below the Gaussian shape coefficient, i.e. 2.

However, before drawing conclusions, remember that, as shown in the previous section, the variance of log-returns is not independent of price levels. Therefore, the i.i.d. assumption does not hold, and the evidence of heavy tails in the dynamics of day-ahead electricity markets may be a statistical artifact due to the mixture of different, possibly non-heavy-tailed processes. In order to account for the effect generated by the presence of heteroskedasticity, we repeat the same Maximum Likelihood estimates of the Subbotin distribution on the rescaled returns defined in (4). Results are reported in Table 3. As it can be seen, the rescaling procedure exerts only a minor impact on the density of log-returns. The heavy-tailed nature of electricity log-returns can be appreciated in Fig. 5.

3 Conclusions

This paper describes some statistical properties of price returns in day-ahead electricity markets. The analysis in [3] is extended to two new markets, namely APX and Powernext, and as such it allows to compare the outcomes of different trading architectures and to assess which properties of the analyzed time series remain the same, and which ones change.

Through the analysis of time dependencies we have shown that the autocorrelation structure of returns in different time series is rather invariant when morning and afternoon (that is, high demand) hours are considered, whereas for low demand hours the short-term autocorrelation seems to slightly vary across markets.

There exists evidence of a negative volatility-price relationship, but only for NordPool and Powernext, and only within the limits of sufficiently low price levels. In days when demand is very high – and so is the market-clearing price – the inverse variance-price relationship breaks down. The existence of a threshold above which the market properties change is consistent with the theoretical results in [10], which characterized the existence of high- and low-demand equilibria with radically different properties, and with regime-switching models of electricity prices, e.g. [11] and [16]. The APX market departs from the observed scaling pattern. Differences in scaling properties may be due to differences in pricing protocols and in marke structures, an issue which deserves deeper investigation.

Finally, concerning the probability distribution of filtered price returns, we can summarize our findings in two points: first, all empirical densities, irre-spectively of the time of the day and of the market considered, are reasonably well described by a Subbotin distribution with shape parameter $b = 1$, that is by a Laplace distribution. Second, the strong presence of heteroskedasticity in the time series does not seem to affect much the estimation of the Subbotin parameters, hence suggesting the relative robustness of these findings.

References

1. Bellini F. (2002) Empirical Analysis of Electricity Spot Prices in European Deregulated Markets. Quaderni Ref. 7/2002.
2. Bystroem H. (2001). Extreme Value Theory and Extremely Large Electricity Price Changes. Mimeo.
3. Bottazzi G., Sapio S., Secchi A. (2005) Some Statistical Investigations on the Nature and Dynamics of Electricity Prices. Physica A 355(1): 54–61.
4. Bottazzi G. (2004) Subbotools: A Reference Manual. LEM Working Paper 2004/14, Scuola Superiore Sant'Anna, Pisa.
5. Bottazzi G., Secchi A. (2003) Why are Distributions of Firm Growth Rates Tent-Shaped?. Economics Letters, 80: 415–420.
6. Bottazzi G., Secchi A. (2006) Maximum Likelihood Estimation of the Sym-metric and Asymmetric Exponential Power Distribution LEM Working Paper 2006/19, Scuola Superiore Sant'Anna, Pisa.
7. Deng S.-J., Jiang W., Xia Z. (2002). Alternative Statistical Specifications of Commodity Price Distribution with Fat Tails. Advanced Modeling and Opti-mization, 4: 1–8.
8. Diebold F.X., Ohanian L.E., Berkovitz J. (1997) Dynamic Equilibrium Economies: A Framework for Comparing Models and Data. Federal Reserve Bank of Philadelphia, working paper No.97-7.
9. Eberlein E., Stahl G. (2003). Both Sides of the Fence: a Statistical and Regula-tory View of Electricity Risk. Energy and Power Risk Management, 8: 34–38.
10. von der Fehr N., Harbord D. (1993). Spot Market Competition in the UK Electricity Industry. Economic Journal 103: 531–546.
11. Huisman R., Mahieu R. (2003). Regime Jumps in Electricity Prices. Energy Economics 25(5): 425–434.
12. Longstaff F.A., Wang A. W. (2002). Electricity Forward Prices: a High-Frequency Empirical Analysis. UCLA Anderson School working paper.
13. Sapio S. (2005) The Nature and Dynamics of Electricity Markets. Doctoral dissertation, Scuola Superiore Sant'Anna, Pisa.
14. Subbotin M.F. (1923). On the Law of Frequency of Errors. Matematicheskii Sbornik, 31: 296–301.
15. Weron R. (2002). Measuring Long-Range Dependence in Electricity Prices. In: Takayasu H. (ed.) Empirical Science of Financial Fluctuations. Springer, Tokyo.
16. Weron R, Bierbrauer M., Truck S. (2004). Modeling Electricity Prices: Jump Diffusion and Regime Switching. Physica A 336: 39–48.

Variations in Financial Time Series: Modelling Through Wavelets and Genetic Programming

Dilip P. Ahalpara[1], Prasanta K. Panigrahi[2], and Jitendra C. Parikh[2]

[1] Institute for Plasma Research, Near Indira Bridge, Gandhinagar-382428, India
   dilip@ipr.res.in
[2] Physical Research Laboratory, Navrangpura, Ahmedabad-380009, India
   prasanta@prl.res.in, parikh@prl.res.in

Summary. We analyze the variations in S&P CNX NSE daily closing index stock values through discrete wavelets. Transients and random high frequency components are effectively isolated from the time series. Subsequently, small scale variations as captured by Daubechies level 3 and 4 wavelet coefficients and modelled by genetic programming. We have smoothened the variations using Spline interpolation method, after which it is found that genetic programming captures the dynamical variations quite well through Padé type of map equations. The low-pass coefficients representing the smooth part of the data has also been modelled. We further study the nature of the temporal variations in the returns.

1 Introduction

Financial time series reveal a variety of temporal variations depending on the time scale of analysis [1–3]. At short time scales one observes predominantly stochastic behavior, as also sharp transients originating from certain physical phenomena [4–6]. In the intermediate time scales, the random variations are averaged out and the possibility of manifestation of structured variations present in the data emerges. Wavelet transform, because of its multi-resolution analysis capability is well suited for this purpose [7–10].

We have studied the time series data of S&P CNX Nifty index of National Stock Exchange (NSE) of India. The daily closing index values are examined for the purpose of identifying the different temporal variations. Daubechies-4 (Db4) wavelet transform has been used for this purpose. The multi-level decomposition clearly reveals stochastic variations at small scales to ordered and cyclic fluctuations at larger time scales [11, 12]. We have found that, wavelet coefficients showed a linear trend in the case of Haar wavelet transform; hence the Db4 higher level coefficients have been studied for modelling purpose.

More precisely, we have used Genetic Programming to model the 3rd, 4th and 5th level Db4 high-pass coefficients. Before carrying out modelling, the coefficients have been interpolated through Spline for the purpose of smoothening. The low-pass coefficients which represent the average data have also been modelled through GP framework. It is observed that GP captures the dynamics of the variations and the average behavior quite well by producing crisp Padẽ type map equations.

2 Wavelet Coefficients

We consider the financial NSE Nifty data set of 2048 points covering the daily closing index corresponding to the duration 24-Dec-1997 to 16-Feb-2006. This daily index is shown in Fig. 1. We then consider Db4 wavelet forward transformation to separate the fluctuations from the trends. The Db4 wavelet transformation is made at different levels ranging from 1 to 8 on the 2048 data set and the resulting wavelet coefficients are shown in Fig. 2(a). One observes structured variations at higher levels, apart from random and transient phenomena at lower scales. The low-pass coefficients at level 8 are shown in Fig. 2(b), which represent the average behavior at this level. The corresponding high-pass and low-pass normalized power has been shown in Figs. 3(a) and 3(b). One clearly sees that at higher scales substantial amount of variation is present, since the low-pass power decreases and the high-pass power increases at higher levels. Considering the structured behavior of the variations, it is interesting to see how well these fluctuation coefficients can be analyzed through the techniques of Genetic Programming in which the model equations are built in the reconstructed phase space.

Fig. 1. NSE Nifty data having 2048 points covering the daily index lying within 24-Dec-1997 to 16-Feb-2006.

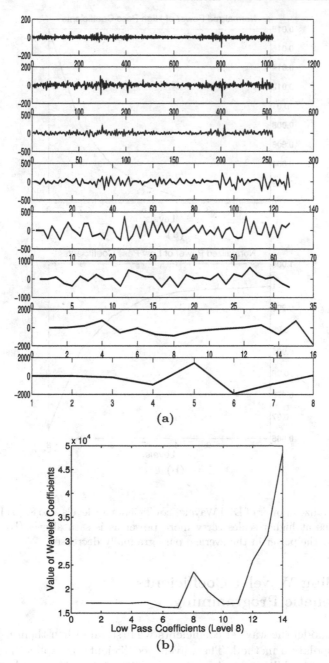

Fig. 2. Db4 Wavelet coefficients, (a) high-pass for levels 1 to 8, (b) low-pass for level 8.

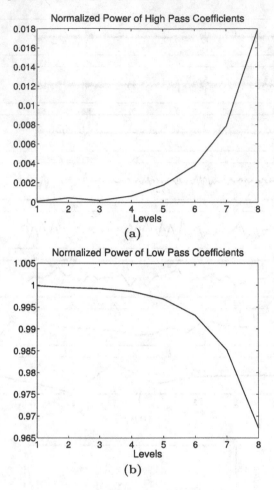

Fig. 3. Normalized power of Db4 Wavelet coefficients for levels 1 to 8. (**a**) High-pass: The variations at higher scales carry more power as is seen above. (**b**) Low-pass: One sees that the power in the average part gradually decreases.

3 Modelling Wavelet Coefficients Using Genetic Programming

In order to model the wavelet coefficients we first smoothen them using cubic Spline interpolation method. The wavelet coefficients are spline interpolated by generating 4 additional points that are sampled within each consecutive pair of points within the data set. The resulting data set is roughly 4 fold bigger and is quite smooth and therefore easy to model. Moreover the piece-wise polynomial form of the interpolated data is well suited with a similar structure used for the map equation, searched by Genetic Programming.

Under the umbrella of Evolutionary Algorithm [13–15] various approaches like Genetic Algorithm [16], Genetic Programming [17] etc, have been used that help solve a variety of optimization problems based on a complex search. In the present note we have incorporated Genetic Programming (GP) that employs a non-linear structure for the chromosomes representing the candidate solutions.

In Genetic Programming we start with an ensemble of chromosomes (called a population) and then create its successive generations stochastically using a set of genetic operators, namely copy, crossover and mutation. A chromosome essentially represents a candidate solution whose fitness is decided by an objective function that maps the chromosome structure to a fitness value. Chromosomes that are having better fitness values are more likely to breed offsprings in the next generation. This evolutionary cycle of selection of chromosomes followed by their transformation using genetic operators leads to the emergence of a population having more and more suitable chromosomes that helps us find a global optimized solution. At the end of a reasonable number of generations, the top chromosome in the population is used as the solution of the optimization problem.

For a given time series X_t, we use the standard embedding technique [18] to reconstruct the phase space using time delayed vectors and carry out a GP fit to generate the map equation of the following form:

$$X_{t+1} = f\left(X_t, X_{t-\tau}, X_{t-2\tau}, \ldots X_{t-(d-1)\tau}\right) \tag{1}$$

where f represents a function involving time series values X_t in the immediate past, arithmatic operators ($+$, $-$, $\times$ and $\div$) and numbers bound between -10 and 10 with a precision of 1 decimal; d represents number of previous time series values that may appear in the function and τ represents a time delay.

The sum of squared errors ($\triangle^2$) is minimized within the GP framework. For a given chromosome, the lower the sum of squared errors, the fitter is the chromosome. The fitness measure is defined as:

$$R^2 = 1 - \frac{\triangle^2}{\sum\limits_{i=1}^{i=N}\left(X_i^{given} - \overline{X_i^{given}}\right)^2} \tag{2}$$

where $\overline{X_i^{given}}$ is the average of all X_t (Eq. (1)) to be fitted.

As described in [17], the Genetic programming is discouraged to overfit by generating longer strings of chromosomes. This is achieved by modifying the fitness measure as follows,

$$r = 1 - \left(1 - R^2\right)\frac{N-1}{N-k} \tag{3}$$

where N is the number of equations to be fitted in the training set and k is the total number of time lagged variables of the form X_t, $X_{t-\tau}$, $X_{t-2\tau}, \ldots$

etc. (including repetitions) occurring in the given chromosome. This modified fitness measure prefers a parsimonious model. It may be noted that for R^2 close to 0, r can be negative.

3.1 Modelling Db4 3rd Level Wavelet Coefficients

The Db4 3rd level forward wavelet coefficients having 254 points are shown in Fig. 4(a). These wavelet coefficients are Spline interpolated to give 1266 data points shown in Fig. 4(b).

In order to search for the map equation, time delayed vectors are used to construct 1250 equations corresponding to dimension = 5 and $\tau = 1$. The resulting map equation generated by GP has a very good fitness value 0.993268 and is shown below, where the numbers are shown with a precision of 4 significant digits for the sake of convenience:

$$X_{t+1} = 1.2404X_t - 0.6667X_{t-\tau} + \frac{0.4545\left(X_t - 2X_{t-2\tau} + 2X_{t-4\tau}\right)}{5.54X_{t-\tau} + 8.8X_{t-2\tau} + 7.84} \tag{4}$$

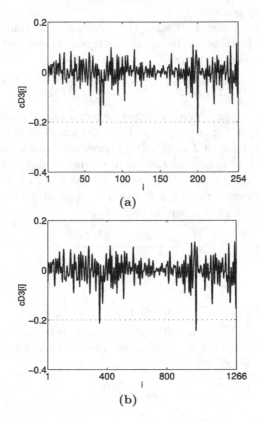

(a)

(b)

Fig. 4. (a) Db4 3rd level wavelet coefficients, (b) Spline interpolated Db4 3rd level wavelet coefficients.

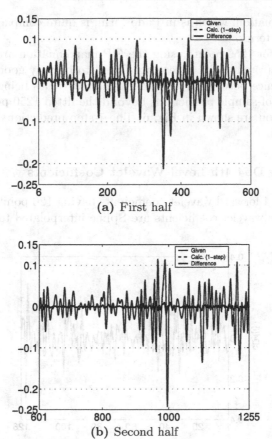

Fig. 5. Fit for 3rd level wavelet coefficients using GP solution.

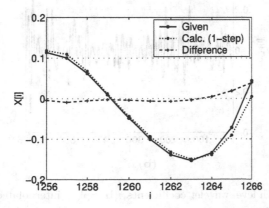

Fig. 6. Out-of-sample 1-step predictions using GP solution for 3rd level wavelet coefficients.

The map equation, which is in Padē form, is quite compact and contains some nonlinear terms.

The GP fit for 1250 points using the GP map equation are shown in two parts in Figs. 5(a) and 5(b). These are found to be very good, as is evident from the difference between given and calculated values being close to zero. The 1-step out-of-sample predictions beyond the fitted 1250 points using the GP map equation are shown in Fig. 6. The 1-step predictions are also found to be very good.

3.2 Modelling Db4 4th Level Wavelet Coefficients

The Db4 level=4 forward wavelet coefficients having 126 points are shown in Fig. 7(a). These wavelet coefficients are Spline interpolated to give 626 data

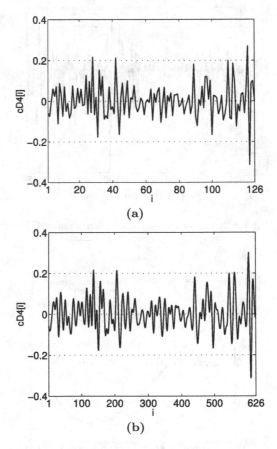

(a)

(b)

Fig. 7. (a) Db4 4th level wavelet coefficients, (b) Spline interpolated Db4 4th level wavelet coefficients.

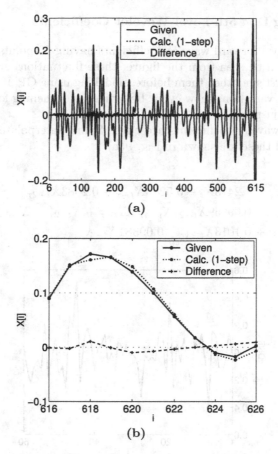

Fig. 8. (a) Fit for 610 data points using GP solution for Db4 4th level wavelet coefficients, (b) Out-of-sample 1-step predictions using GP solution for Db4 4th level wavelet coefficients.

points and are shown in Fig. 7(b). The map equation generated by GP for dimension = 5 and $\tau = 1$ has fitness value 0.9984 and is shown below:

$$X_{t+1} = 2.14X_t - 0.54\left(1.1 - X_{t-\tau}\right)\left(X_t - X_{t-\tau}\right)\left(X_t - X_{t-3\tau}\right) - 2.58X_{t-\tau}$$

$$+1.08X_{t-2\tau} + 0.27X_{t-3\tau} - 0.27X_{t-4\tau} \tag{5}$$

The model map equation is quite compact and contains some nonlinear terms. The GP fit for 610 points using the GP map equation is shown in Fig. 8(a) and is found to be very good.

The 1-step out-of-sample predictions beyond the fitted 610 points using the GP map equation are shown in Fig. 8(b). The 1-step predictions are found to be very good.

3.3 Modelling Db4 5th Level Wavelet Coefficients

The Db4 5th level forward wavelet coefficients having 60 points are shown in Fig. 9(a). As can be seen from the figure, these fluctuations are not smooth and hence we first smoothen them before modelling using GP. It may be noted that since these variations show cyclic behavior, smoothening for the purpose of modelling is justified.

The above wavelet coefficients are cubic Spline interpolated to give 296 data points and these are shown in Fig. 9(b).

$$
\begin{aligned}
X_{t+1} = {} & 1.2473X_t - 0.5283X_{t-\tau} - 0.2983X_{t-2\tau} \\
& + 0.09669X_{t-2\tau}X_{t-4\tau}\left(X_{t-\tau} + X_{t-2\tau} + X_{t-3\tau}\right) \\
& + 0.1013X_{t-3\tau} + 0.09381X_{t-4\tau}
\end{aligned}
\tag{6}
$$

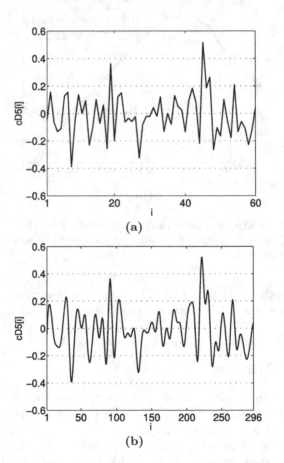

(a)

(b)

Fig. 9. (a) Db4 5th level wavelet coefficients, (b) Spline interpolated Db4 5th level wavelet coefficients.

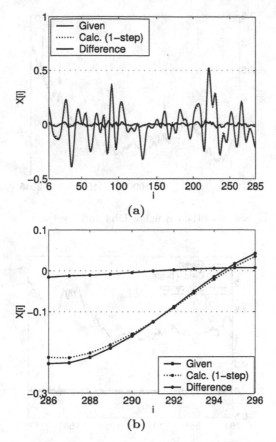

(a)

(b)

Fig. 10. (a) Fit for 280 data points using GP solution for Db4 level-5 wavelet coefficients, (b) Out-of-sample 1-step predictions using GP solution for Db4 level-5 wavelet coefficients.

The map equation generated by GP for dimension = 5 and $\tau = 1$ is having fitness value 0.992478 and is shown in Eq. (6). Interestingly the map equation is primarily linear with a small component of non-linear term. The GP fit for 280 points using the GP map equation is shown in Fig. 10(a). The thick line in the figure close to 0.0 indicates that the differences between calculated and given values are very small and hence the fit is quite good. The map equation is then used to make out-of-sample predictions. The 1-step out-of-sample predictions beyond 280 points using the GP map equation are shown in Fig. 10(b).

3.4 Modelling Db4 2nd Level Average Wavelet Coefficients

We have also considered GP analysis on the average (trend) wavelet coefficients for S&P CNX NSE daily closing index stock values. This is first smoothened using a Db4 2nd level transform and is shown in Fig. 11.

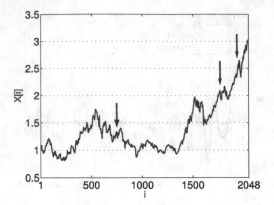

Fig. 11. NSE data smoothened using Db4 2nd level wavelet transform.

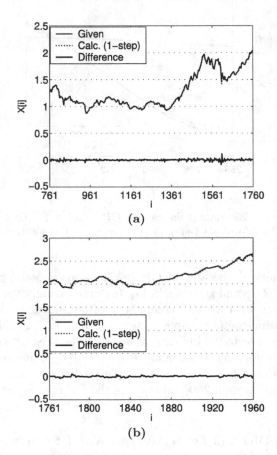

Fig. 12. (a) GP fit of 1000 points for Db4 2nd level smoothened NSE data set, (b) Out-of-sample predictions for 200 points beyond fitted region of 1000 points for Db4 2nd level smoothened NSE data set.

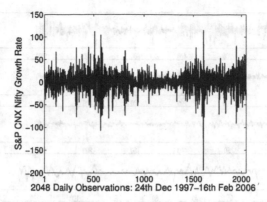

Fig. 13. S&P CNX Nifty returns, having 2048 points covering the daily index lying within 24-Dec-1997 to 16-Feb-2006.

With reference to Fig. 11, we consider a total of 1200 points in which initial 1000 points (lying within point no. 745 and 1744 and marked by first two arrows) are used for fitting and further 200 out-of-sample points (upto point no. 1944 also marked by arrow in the figure) are used for predictions.

The GP fit for 1000 points is shown in Fig. 12(a). It gives a fitness values of 0.9985 with a Padẽ form solution as shown in Eq. (7).

$$X_{t+1} = X_t^N / X_t^D$$
$$X_t^N = X_{t4} \left(X_{t1} * X_{t2} * X_{t4}^2 + X_{t1}^2 * X_{t2}^2 + X_{t1} * X_{t4}^3 + X_{t1}^3 * X_{t4} \right.$$
$$\left. -X_{t1}^4 - X_{t2} * X_{t4}^3 - X_{t1}^2 * X_{t2} * X_{t4} + X_{t1}^3 * X_{t2} \right)$$
$$X_t^D = X_{t1} \left(X_{t2} * X_{t4}^2 + X_{t1} * X_{t2}^2 + X_{t4}^3 + X_{t1}^2 * X_{t4} - X_{t1}^3 \right) \qquad (7)$$

For the sake of curiosity, we have also studied the nature of the returns in the above data set. Fig. 13 shows these returns, whose Haar wavelet coefficients are shown in Fig. 14(a). One observes transients and cyclic behavior at various levels. The 8th level low-pass coefficients are shown in Fig. 14(b). The linear trend in the cyclic variations of the returns is clearly visible. In this light it may be interesting to study the predictability of the higher level cyclic behavior. The same has been attempted through GP framework with substantial success [19].

4 Conclusion

In conclusion, the multi-resolution analysis of discrete wavelet transforms can be effectively combined with Genetic Programming for the purpose of predicting small scale structured variations. It is expected that higher scale cyclic variations can be effectively modelled. Keeping in mind the stochastic nature of the very small scale fluctuations, this approach opens up the possibility of

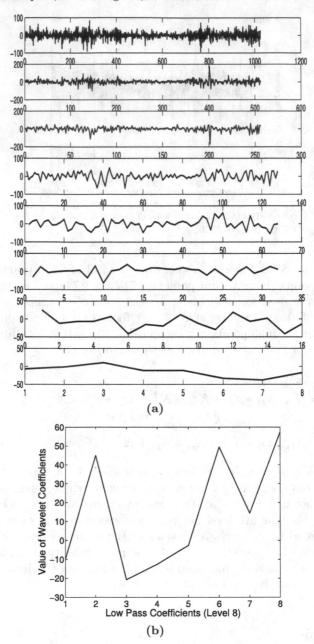

(a)

(b)

Fig. 14. Haar Wavelet coefficients for the return time series, (**a**) high-pass for levels 1 to 8, (**b**) low-pass for level 8.

modelling financial time series in the medium time scales. The smooth part of the data or the trend has been modelled and reported in [20] and has been described in detail in [21].

References

1. Mandelbrot BB (1997) Multifractals and 1/f Noise. Springer-Verlag, New York
2. Mantegna RN, Stanley HE (2000) An Introduction to Econophysics: Correlations and Complexity in Finance. Cambridge University Press
3. Clark PK (1973) A Subordinated Stochastic Process Model with Finite Variance for Speculative Prices. Econometrica, 41:135–256
4. Mankiw MG, Romer D, Shapiro MD (1997) Stock Market forecastability and volatility: a statistical appraisal, Review of Economic Studies, 58:455–77
5. Grassberger P, Procaccia I (1983) Characterization of Strange Attractors, Physical Review Letters, 50:346–49
6. Poon S-H, Granger CWJ (2003) Forecasting Volatility in Financial Markets: A Review, Journal of Economic Literature, 41:478–539
7. Daubechies I (1992) Ten Lectures on Wavelets, Vol. 64 of CBMS-NSF Regional Conference Series in Applied Mathematics. Society of Industrial and Applied Mathematics, Philadelphia
8. Gencay R, Selcuk F, Whitcher B (2001) An Introduction to Wavelets and Other Filtering Methods in Finance and Economics. Academic Press
9. Connor J, Rossiter R (2005) Wavelet Transforms and Commodity Prices, Studies in Nonlinear Dynamics and Econometrics, 9:1–20
10. Percival D, Walden A (2000) Wavelet Analysis for Time Series Analysis. Cambridge University Press
11. Ramsey JB, Usikov D, Zaslavsky G (1995) An Analysis of US Stock Price Behavior Using Wavelets, Fractals, 3:377–89
12. Biswal PC, Kamaiah B, Panigrahi PK (2004) Wavelet Analysis of the Bombay Stock Exchange Index, Journal of Quantitative Economics 2:133–46
13. Holland JH (1975) Adaptation in Natural and Artifical Systems. University of Michigan Press, Ann Arbor 2nd ed.
14. Goldberg DE (1989) Genetic Algorithms in Search, Optimization, and Machine Learning. Addison Wesley publication
15. Fogel DB (1998) Evolutionary Computation, The Fossil Record. IEEE Press
16. Mitchell M (1996) An Introduction to Genetic Algorithms. MIT Press
17. Szpiro GG (1997) Forecasting chaotic time series with genetic algorithms, Physical Review E, 55:2557–2568
18. Abarbanel H, Brown R, Sidorovich JJ, Tsimring L (1993) The analysis of observed chaotic data in physical systems , Rev. of Mod. Phys., 65:1331–1392
19. Ahalpara DP, Verma A, Panigrahi PK, Parikh JC (2007) Characterizing and modeling cyclic behavior in non-stationary time series through multi-resolution analysis, submitted to The Euro. Phys. J. B
20. Manimaran PM, Parikh JC, Panigrahi PK, Basu S, Kishtewal CM, Porecha MB (2006) Modelling Financial Time Series. In: Chatterjee A, Chakrabarti BK (eds)Econophysics of Stock and Other Markets. Springer-Verlag, Italy
21. Ahalpara DP, Parikh JC (2007) submitted for publication in Physica A

Financial Time-series Analysis: a Brief Overview

A. Chakraborti[1], M. Patriarca[2], and M.S. Santhanam[3]

[1] Department of Physics, Banaras Hindu University, Varanasi-221 005, India
`achakraborti@yahoo.com`
[2] Institute of Theoretical Physics, Tartu University, Tähe 4, 51010 Tartu, Estonia
`marco.patriarca@mac.com`
[3] Physical Research Laboratory, Navrangpura, Ahmedabad-380 009, India
`santh@prl.ernet.in`

1 Introduction

Prices of commodities or assets produce what is called time-series. Different kinds of financial time-series have been recorded and studied for decades. Nowadays, all transactions on a financial market are recorded, leading to a huge amount of data available, either for free in the Internet or commercially. Financial time-series analysis is of great interest to practitioners as well as to theoreticians, for making inferences and predictions. Furthermore, the stochastic uncertainties inherent in financial time-series and the theory needed to deal with them make the subject especially interesting not only to economists, but also to statisticians and physicists [1]. While it would be a formidable task to make an exhaustive review on the topic, with this review we try to give a flavor of some of its aspects.

2 Stochastic Methods in Time-series Analysis

The birth of physics as a science is usually associated with the study of mechanical objects moving with negligible fluctuations, such as the motion of planets. However, this type of systems is not unique, especially at smaller scales where the interaction with the environment and its influence in the form of random fluctuations has to be taken into account. The main theoretical tool to describe the evolution of such systems is the theory of stochastic processes, which can be formulated in various ways: in terms of a Master equation, Fokker-Planck type equation, random walk model, Langevin equation, or through path integrals. Some systems can present unpredictable chaotic behavior due to dynamically generated internal noise. Either truly stochastic or chaotic in nature, noisy processes represent the rule rather than an exception, not only in condensed matter physics but in many fields such as cosmology, geology, meteorology, ecology, genetics, sociology, and economics. In fact the first formulation of the random walk model and a stochastic process was given in the framework of an economic study [2, 3]. In the following we propose and discuss some questions which we consider as possible land-marks in the field of time series analysis.

2.1 Time-series Versus Random Walk

What if the time-series were similar to a random walk? The answer is: It would not be possible to predict future price movements using the past price movements or trends. Louis Bachelier, who was the first one to investigate this issue in 1900 [2], reached the conclusion that "The mathematical expectation of the speculator is zero" and described this condition as a "fair game."

In economics, if $P(t)$ is the price of a stock or commodity at time t, then the "log-return" is defined as $r_\tau(t) = \ln P(t + \tau) - \ln P(t)$, where τ is the interval of time. Some statistical features of daily log-return are illustrated in Fig. 1, using the price time-series for the General Electric. The

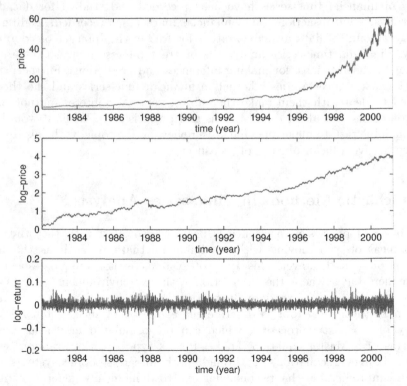

Fig. 1. Price in USD (*above*), log-price (*center*) and log-return (*below*) plotted versus time for the General Electric during the period 1982–2000.

real empirical returns are compared in Fig. 2 with a random time-series
we generated using random numbers extracted from a Normal distribution
with zero mean and unit standard deviation. If we divide the time-interval
τ into N sub-intervals (of width Δt), the total log-return $r_\tau(t)$ is by definition the sum of the log-returns in each sub-interval. If the price changes in
each sub-interval are independent (Fig. 2 above) and identically distributed
with a finite variance, according to the central limit theorem the cumulative
distribution function $F(r_\tau)$ converges to a Gaussian (Normal) distribution
for large τ. The Gaussian (Normal) distribution has the following properties: (a) the average and most probable change is zero; (b) the probability
of large fluctuations is very low; (c) it is a *stable* distribution. The distribution of returns was first modeled for "bonds" [2] as a Normal distribution,

$$P(r) = \left[\sqrt{2\pi}\sigma\right]^{-1} \exp\left(-r^2/2\sigma^2\right),$$

where σ^2 is the variance of the distribution.

In the classical financial theories Normality had always been assumed,
until Mandelbrot [4] and Fama [5] pointed out that the empirical return distributions are fundamentally different. Namely, they are "fat-tailed" and more
peaked compared to the Normal distribution. Based on daily prices in different
markets, Mandelbrot and Fama found that $F(r_\tau)$ was a stable Levy distribution whose tail decays with an exponent $\alpha \simeq 1.7$. This result suggested that
short-term price changes were not well-behaved since most statistical prop-

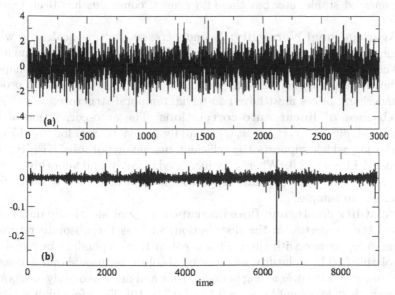

Fig. 2. Random time-series, 3000 time steps (*above*) and Return time-series of the
S&P500 stock index, 8938 time steps (*below*).

erties are not defined when the variance does not exist. Later, using more extensive data, the decay of the distribution was shown to be fast enough to provide finite second moment. With time, several other interesting features of the financial data were unearthed.

The motive of physicists in analyzing financial data has been to find common or universal regularities in the complex time-series (a different approach from those of the economists doing traditional statistical analysis of financial data). The results of their empirical studies on asset price series show that the apparently random variations of asset prices share some statistical properties which are interesting, non-trivial, and common for various assets, markets, and time periods. These are called "stylized empirical facts".

2.2 "Stylized" Facts

Stylized facts are usually formulated using general *qualitative* properties of asset returns. Hence, distinctive characteristics of the individual assets are not taken into account. Below we consider a few ones from [6].

(i) **Fat tails**: Large returns asymptotically follow a power law $F(r_\tau) \sim |r|^{-\alpha}$, with $\alpha > 2$. The values $\alpha = 3.01 \pm 0.03$ and $\alpha = 2.84 \pm 0.12$ are found for the positive and negative tail respectively [7]. An $\alpha > 2$ ensures a well-defined second moment and excludes stable laws with infinite variance. There have been various suggestions for the form of the distribution: Student's-t (Fig. 3), hyperbolic, normal inverse Gaussian, exponentially truncated stable, etc. but there no general consensus has been reached yet

(ii) **Aggregational Normality**: As one increases the time scale over which the returns are calculated, their distribution approaches the Normal form. The shape is different at different time scales. The fact that the shape of the distribution changes with τ makes it clear that the random process underlying prices must have non-trivial temporal structure.

(iii) **Absence of linear auto-correlations**: The auto-correlation of log-returns, $\rho(T) \sim \langle r_\tau(t+T)r_\tau(t) \rangle$, rapidly decays to zero for $\tau \geq 15$ minutes [8], which supports the "efficient market hypothesis" (EMH), discussed in Sect. 2.3. When τ is increased, weekly and monthly returns exhibit some auto-correlation but the statistical evidence varies from sample to sample.

(iv) **Volatility clustering**: Price fluctuations are not identically distributed and the properties of the distribution, such as the absolute return or variance, change with time. This is called time-dependent or "clustered volatility". The volatility measure of absolute returns shows a positive auto-correlation over a long period of time and decays roughly as a power-law with an exponent between 0.1 and 0.3 [8–10]. Therefore high volatility events tend to cluster in time, large changes tend to be followed by large changes, and analogously for small changes.

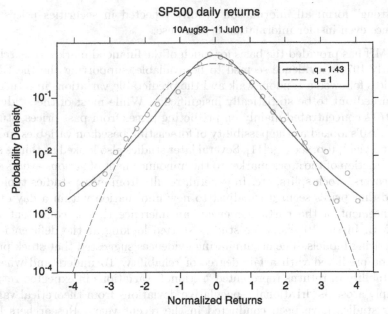

Fig. 3. S&P 500 daily return distribution and normal kernel density estimate. Distributions of log returns normalized by the sample standard deviation rising from the demeaned S & P 500 (*circles*) and from a Tsallis distribution of index $q = 1.43$ (*solid line*). For comparison, the normal distribution $q = 1$ is shown (*dashed line*). Reprinted and adapted from L. Borland, *A Theory of Non-Gaussian Option Pricing*, arxiv:cond-mat/0205078; original figure in L.B. Robert Osorio and C. Tsallis, *Distributions of High-Frequency Stock-Market Observables*, in: *Nonextensive Entropy: Interdisciplinary Applications*, M. Gell-Mann and C. Tsallis Eds., Santa Fe Institute Studies on the Sciences of Complexity, Oxford University Press, Oxford, 2004; with permission from Oxford University Press.

2.3 The Efficient Market Hypothesis (EMH)

A debatable issue in financial econometrics is whether the market is "efficient" or not. The "efficient" asset market is that in which the information contained in past prices is instantly, fully and continually reflected in the asset's current price. The EMH was proposed by Eugene Fama in his Ph.D. thesis work in the 1960's, in which he argued that in an active market that consists of intelligent and well-informed investors, securities would be fairly priced and reflect all the information available. Till date there continues to be disagreement on the degree of market efficiency. The three widely accepted forms of the EMH are:

- "Weak" form: all past market prices and data are fully reflected in securities prices and hence technical analysis is of no use.
- "Semistrong" form: all publicly available information is fully reflected in securities prices and hence fundamental analysis is of no use.

- "Strong" form: all information is fully reflected in securities prices and hence even insider information is of no use.

The EMH has provided the basis for much of the financial market research. In the early 1970's, evidence seemed to be available, supporting the the EMH: the prices followed a random walk and the predictable variations in returns, if any, turned out to be statistically insignificant. While most of the studies in the 1970's concentrated mainly on predicting prices from past prices, studies in the 1980's looked at the possibility of forecasting based on variables such as dividend yield, too, see e.g. [11]. Several later studies also looked at things such as the reaction of the stock market to the announcement of various events such as takeovers, stock splits, etc. In general, results from event studies typically showed that prices seemed to adjust to new information within a day of the announcement of the particular event, an inference that is consistent with the EMH. In the 1990's, some studies started looking at the deficiencies of asset pricing models. The accumulating evidences suggested that stock prices could be predicted with a fair degree of reliability. To understand whether predictability of returns represented "rational" variations in expected returns or simply arose as "irrational" speculative deviations from theoretical values, further studies have been conducted in the recent years. Researchers have now discovered several stock market "anomalies" that seem to contradict the EMH. Once an anomaly is discovered, in principle, investors attempting to profit by exploiting such an inefficiency should result in the disappearance of the anomaly. In fact, many such anomalies that have been discovered via back-testing, have subsequently disappeared or proved to be impossible to exploit due to high costs of transactions.

We would like to mention the paradoxical nature of efficient markets: if every practitioner truly believed that a market was efficient, then the market would not have been efficient since no one would have then analyzed the behavior of the asset prices. In fact, efficient markets depend on market participants who believe the market is inefficient and trade assets in order to make the most of the market inefficiency.

2.4 Are There Any Long-time Correlations?

Two of the most important and simple models of probability theory and financial econometrics are the random walk and the Martingale theory. They assume that the future price changes only depend on the past price changes. Their main characteristic is that the returns are uncorrelated. But are they truly uncorrelated or are there long-time correlations in the financial time-series? This question has been studied especially since it may lead to deeper insights about the underlying processes that generate the time-series [12].

Next we discuss two measures to quantify the long-time correlations, and study the strength of trends: the R/S analysis to calculate the Hurst exponent and the detrended fluctuation analysis [13].

Hurst Exponent from R/S Analysis

In order to measure the strength of trends or "persistence" in different processes, the rescaled range (R/S) analysis to calculate the Hurst exponent can be used. One studies the rate of change of the rescaled range with the change of the length of time over which measurements are made. We divide the time-series ξ_t of length T into N periods of length τ, such that $N\tau = T$. For each period $i = 1, 2, \ldots, N$, containing τ observations, the cumulative deviation is

$$X(\tau) = \sum_{t=(i-1)\tau+1}^{i\tau} (\xi_t - \langle \xi \rangle_\tau), \tag{1}$$

where $\langle \xi \rangle_\tau$ is the mean within the time-period and is given by

$$\langle \xi \rangle_\tau = \frac{1}{\tau} \sum_{t=(i-1)\tau+1}^{i\tau} \xi_t. \tag{2}$$

The range in the i-th time period is given by $R(\tau) = \max X(\tau) - \min X(\tau)$, and the standard deviation is given by

$$S(\tau) = \left[\frac{1}{\tau} \sum_{t=(i-1)\tau+1}^{i\tau} (\xi_t - \langle \xi \rangle_\tau)^2 \right]^{\frac{1}{2}}. \tag{3}$$

Then $R(\tau)/S(\tau)$ is asymptotically given by a power-law

$$R(\tau)/S(\tau) = \kappa \tau^H, \tag{4}$$

where κ is a constant and H the Hurst exponent. In general, "persistent" behavior with fractal properties is characterized by a Hurst exponent $0.5 < H \leq 1$, random behavior by $H = 0.5$ and "anti-persistent" behavior by $0 \leq H < 0.5$. Usually Eq. (4) is rewritten in terms of logarithms, $\log(R/S) = H \log(\tau) + \log(\kappa)$, and the Hurst exponent is determined from the slope.

Detrended Fluctuation Analysis (DFA)

In the DFA method the time-series ξ_t of length T is first divided into N non-overlapping periods of length τ, such that $N\tau = T$. In each period $i = 1, 2, \ldots, N$ the time-series is first fitted through a linear function $z_t = at + b$, called the local trend. Then it is detrended by subtracting the local trend, in order to compute the fluctuation function,

$$F(\tau) = \left[\frac{1}{\tau} \sum_{t=(i-1)\tau+1}^{i\tau} (\xi_t - z_t)^2 \right]^{\frac{1}{2}}. \tag{5}$$

The function $F(\tau)$ is re-computed for different box sizes τ (different scales) to obtain the relationship between $F(\tau)$ and τ. A power-law relation between $F(\tau)$ and the box size τ, $F(\tau) \sim \tau^\alpha$, indicates the presence of scaling. The scaling or "correlation exponent" α quantifies the correlation properties of the signal: if $\alpha = 0.5$ the signal is uncorrelated (white noise); if $\alpha > 0.5$ the signal is anti-correlated; if $\alpha < 0.5$, there are positive correlations in the signal.

Comparison of Different Time-series

Besides comparing empirical financial time-series with randomly generated time-series, here we make the comparison with multivariate spatiotemporal time-series drawn from coupled map lattices and the multiplicative stochastic process GARCH(1,1) used to model financial time-series.

Multivariate Spatiotemporal Time-series Drawn from Coupled Map Lattices

The concept of coupled map lattices (CML) was introduced as a simple model capable of displaying complex dynamical behavior generic to many spatiotemporal systems [14,15]. Coupled map lattices are discrete in time and space, but have a continuous state space. By changing the system parameters, one can tune the dynamics toward the desired spatial correlation properties, many of them already studied and reported [15]. We consider the class of diffusively coupled map lattices in one-dimension, with sites $i = 1, 2, \ldots, n$, of the form

$$y_{t+1}^i = (1 - \epsilon)f\left(y_t^i\right) + \epsilon \left[\, f\left(y_t^{i+1}\right) + f\left(y_t^{i-1}\right)\,\right]/2\,, \qquad (6)$$

where $f(y) = 1 - a\,y^2$ is the logistic map whose dynamics is controlled by the parameter a and the parameter ϵ measures the coupling strength between nearest-neighbor sites. We generally choose periodic boundary conditions, $x(n+1) = x(1)$. In the numerical computations reported by Chakraborti and Santhanam [16], a coupled map lattice with $n = 500$ was iterated starting from random initial conditions, for $p = 5 \times 10^7$ time steps, after discarding 10^5 transient iterates. As the parameters a and ϵ are varied, the spatiotemporal map displays various dynamical features like frozen random patterns, pattern selection, space-time intermittency, and spatiotemporal chaos [15]. In order to study the coupled map lattice dynamics found in the regime of spatiotemporal chaos, where correlations are known to decay rather quickly as a function of the lattice site, the parameters were chosen as $a = 1.97$ and $\epsilon = 0.4$.

Multiplicative Stochastic Process GARCH(1,1)

Considerable interest has been in the application of ARCH/GARCH models to financial time-series, which exhibit periods of unusually large volatility followed by periods of relative tranquility. The assumption of constant variance or "homoskedasticity" is inappropriate in such circumstances. A stochastic process with auto-regressional conditional "heteroskedasticity" (ARCH) is actually a stochastic process with "non-constant variances conditional on the past but constant unconditional variances" [17]. The ARCH(p) process is defined by the equation

$$\sigma_t^2 = \alpha_0 + \alpha_1 x_{t-1}^2 + \ldots + \alpha_p x_{t-p}^2, \tag{7}$$

where the $\{\alpha_0, \alpha_1, \ldots \alpha_p\}$ are positive parameters and x_t is a random variable with zero mean and variance σ_t^2, characterized by a conditional probability distribution function $f_t(x)$, which may be chosen as Gaussian. The nature of the memory of the variance σ_t^2 is determined by the parameter p.

The generalized ARCH process GARCH(p, q) was introduced by Bollerslev [18] and is defined by the equation

$$\sigma_t^2 = \alpha_0 + \alpha_1 x_{t-1}^2 + \ldots + \alpha_q x_{t-q}^2 + \beta_1 \sigma_{t-1}^2 + \ldots + \beta_p \sigma_{t-p}^2, \tag{8}$$

where $\{\beta_1, \ldots, \beta_p\}$ are additional control parameters.

The simplest GARCH process is the GARCH(1,1) process, with Gaussian conditional probability distribution function,

$$\sigma_t^2 = \alpha_0 + \alpha_1 x_{t-1}^2 + \beta_1 \sigma_{t-1}^2. \tag{9}$$

The random variable x_t can be written in term of σ_t defining $x_t \equiv \eta_t \sigma_t$, where η_t is a random Gaussian process with zero mean and unit variance. One can rewrite Eq. 9 as a random multiplicative process

$$\sigma_t^2 = \alpha_0 + (\alpha_1 \eta_{t-1}^2 + \beta_1) \sigma_{t-1}^2. \tag{10}$$

DFA Analysis of Auto-correlation Function of Absolute Returns

The analysis of financial correlations was done in 1997 by the group of H.E. Stanley [9]. The correlation function of the financial indices of the New York stock exchange and the S&P 500 between January, 1984 and December, 1996 were analyzed at one minute intervals. The study confirmed that the auto-correlation function of the returns fell off exponentially but the absolute value of the returns did not. Correlations of the absolute values of the index returns could be described through two different power laws, with crossover time $t_\times \approx 600$ minutes, corresponding to 1.5 trading days. Results from power spectrum analysis and DFA analysis were found to be consistent. The power spectrum analysis of Fig. 4 yielded exponents $\beta_1 = 0.31$ and $\beta_2 = 0.90$ for $f > f_\times$ and $f < f_\times$, respectively. This is consistent with the result that $\alpha = (1 + \beta)/2$ and $t_\times \approx 1/f_\times$, as obtained from detrended fluctuation analysis with exponents $\alpha_1 = 0.66$ and $\alpha_2 = 0.93$ for $t < t_\times$ and $t > t_\times$, respectively.

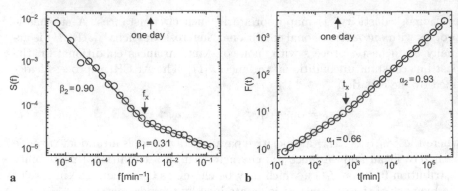

Fig. 4. Power spectrum analysis (*left*) and detrended fluctuation analysis (*right*) of auto-correlation function of absolute returns. Reprinted from Y. Liu, P. Cizeau, M. Meyer, C.-K. Peng, and H.E. Stanley, *Correlations in Economic Time Series*, Physica A **245**, 437 (1997), Copyright (1997), with permission from Elsevier.

Numerical Comparison

In order to provide an illustrative example, in Fig. 5 a comparison among various analysis techniques and process is presented, while the values of the exponents of the Hurst and DFA analyzes are listed in Table 1. For the numerical computations reported by Chakraborti and Santhanam [16], the parameter values chosen were $\alpha_0 = 0.00023$, $\alpha_1 = 0.09$ and $\beta_0 = 0.01$.

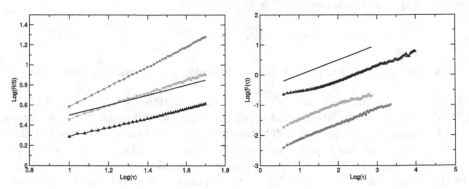

Fig. 5. R/S (*left*) and DFA (*right*) analyses: Random time-series, 3000 time steps (*black solid line*); multivariate spatiotemporal time-series drawn from coupled map lattices with parameters $a = 1.97$ and $\epsilon = 0.4$, 3000 time steps (*black filled up-triangles*); multiplicative stochastic process GARCH(1,1) with parameters $\alpha_0 = 0.00023$, $\alpha_1 = 0.09$ and $\beta_0 = 0.01$, 3000 time steps (*filled squares*); Return time-series of the S&P500 stock index, 8938 time steps (*filled circles*).

Table 1. Hurst and DFA exponents.

Process	Hurst exponent	DFA exponent
Random	0.50	0.50
Chaotic (CML)	0.46	0.48
GARCH(1,1)	0.63	0.51
Financial Returns	0.99	0.51

3 Random Matrix Methods in Time-series Analysis

The R/S and the detrended fluctuation analysis considered in the previous section are suitable for analyzing univariate data. Since the stock-market data are essentially *multivariate* time-series data, it is worth constructing a correlation matrix to study its spectra and contrasting it with random multivariate data from coupled map lattice. Empirical spectra of correlation matrices, drawn from time-series data, are known to follow mostly random matrix theory (RMT) [19].

3.1 Correlation Matrix and Eigenvalue Density

Correlation Matrix

Financial Correlation Matrix

If there are N assets with a price $P_i(t)$ for asset i at time t, the logarithmic return of stock i is $r_i(t) = \ln P_i(t) - \ln P_i(t-1)$. A sequence of such values for a give period of time forms the return vector r_i. In order to characterize the synchronous time evolution of stocks, one defines the equal time correlation coefficients between stocks i and j,

$$\rho_{ij} = [\langle r_i r_j \rangle - \langle r_i \rangle \langle r_j \rangle] / \sqrt{\left[\langle r_i^2 \rangle - \langle r_i \rangle^2\right]\left[\langle r_j^2 \rangle - \langle r_j \rangle^2\right]}, \qquad (11)$$

where $\langle \ldots \rangle$ indicates a time average over the trading days included in the return vectors. The correlation coefficients ρ_{ij} form an $N \times N$ matrix, with $-1 \leq \rho_{ij} \leq 1$. If $\rho_{ij} = 1$, the stock price changes are completely correlated; if $\rho_{ij} = 0$, the stock price changes are uncorrelated and if $\rho_{ij} = -1$, then the stock price changes are completely anti-correlated.

Correlation Matrix from Spatiotemporal Series from Coupled Map Lattices

Consider a time-series of the form $z'(x,t)$, where $x = 1, 2, \ldots, n$ and $t = 1, 2, \ldots, p$ denote the discretized space and time. In this way, the time-series at every spatial point is treated as a different variable. We define

$$z(x,t) = \left[z'(x,t) - \langle z'(x) \rangle \right] / \sigma(x), \tag{12}$$

as the normalized variable, with the brackets $\langle . \rangle$ representing a temporal average and $\sigma(x)$ the standard deviation of z' at position x. Then, the equal-time cross-correlation matrix can be written as

$$S_{x,x'} = \langle z(x,t) \, z(x',t) \rangle, \qquad x, x' = 1, 2, \ldots, n. \tag{13}$$

This correlation matrix is symmetric by construction. In addition, a large class of processes is translationally invariant and the correlation matrix will possess the corresponding symmetry. We use this property for our correlation models in the context of coupled map lattices. In time-series analysis, the averages $\langle . \rangle$ have to be replaced by estimates obtained from finite samples. We use the maximum likelihood estimates, i.e., $\langle a(t) \rangle \approx \frac{1}{p} \sum_{t=1}^{p} a(t)$. These estimates contain statistical uncertainties which disappear for $p \to \infty$. Ideally we require $p \gg n$ to have reasonably correct correlation estimates.

Eigenvalue Density

The interpretation of the spectra of empirical correlation matrices should be done carefully in order to distinguish between system specific signatures and universal features. The former ones express themselves in a smoothed level density, whereas the latter ones are usually represented by the fluctuations on top of such a smooth curve. In time-series analysis, matrix elements are not only prone to uncertainties such as measurement noise on the time-series data, but also to the statistical fluctuations due to finite sample effects. When characterizing time series data in terms of RMT, we are not interested in these sources of fluctuations, which are present on every data set, but we want to identify the significant features which would be shared, in principle, by an "infinite" amount of data without measurement noise. The eigenfunctions of the correlation matrices constructed from such empirical time-series carry the information contained in the original time-series data in a "graded" manner and provide a compact representation for it. Thus, by applying an approach based on RMT, we try to identify non-random components of the correlation matrix spectra as deviations from RMT predictions [19].

We now consider the eigenvalue density, studied in applications of RMT methods to time-series correlations. Let $\mathcal{N}(\lambda)$ be the integrated eigenvalue density, giving the number of eigenvalues smaller than a given λ. The eigenvalue or level density, $\rho(\lambda) = d\mathcal{N}(\lambda)/d\lambda$, can be obtained assuming a random correlation matrix [20]. Results are found to be in good agreement with the empirical time-series data from stock market fluctuations [21]. From RMT considerations, the eigenvalue density for random correlations is given by

$$\rho_{rmt}(\lambda) = [Q/(2\pi\lambda)] \sqrt{(\lambda_{max} - \lambda)(\lambda - \lambda_{min})}. \tag{14}$$

Here $Q = N/T$ is the ratio of the number of variables to the length of each time-series, while $\lambda_{min} = 1 + 1/Q - 2\sqrt{1/Q}$ and $\lambda_{max} = 1 + 1/Q + 2\sqrt{1/Q}$

represent the minimum and maximum eigenvalues of the random correlation matrix. The presence of correlations in the empirical correlation matrix produces a violation of this form of eigenvalue density, for a certain number of dominant eigenvalues, often corresponding to system specific information in the data. As examples, Fig. 6 shows the eigenvalue densities for S&P500 data and for the chaotic data from coupled map lattice are shown: the curves are qualitatively different from the form of Eq. (14).

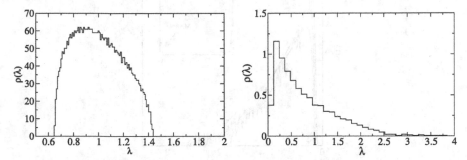

Fig. 6. Spectral density for multivariate spatiotemporal time-series drawn from coupled map lattices (*left*) and eigenvalue density for the return time-series of the S&P500 stock market data, 8938 time steps (*right*).

3.2 Earlier Estimates and Studies Using Random Matrix Theory (RMT)

Laloux et al. [22] showed that results from RMT were useful to understand the statistical structure of the empirical correlation matrices appearing in the study of price fluctuations. The empirical determination of a correlation matrix is a difficult task. If one considers N assets, the correlation matrix contains $N(N-1)/2$ mathematically independent elements, which must be determined from N time-series of length T. If T is not very large compared to N, then generally the determination of the covariances is noisy, and therefore the empirical correlation matrix is to a large extent random. The smallest eigenvalues of the matrix are the most sensitive to this "noise". But the eigenvectors corresponding to these smallest eigenvalues determine the minimum risk portfolios in Markowitz's theory. It is thus important to distinguish "signal" from "noise" or, in other words, to extract the eigenvectors and eigenvalues of the correlation matrix, containing real information (which is important for risk control), from those which do not contain any useful information and are unstable in time. It is useful to compare the properties of an empirical correlation matrix to a "null hypothesis" – a random matrix which arises for example from a finite time-series of strictly uncorrelated assets. Deviations from the random matrix case might then suggest the presence of true information. The main result of the study was a remarkable agreement between theoretical predictions, based on the assumption that the correlation matrix is random, and empirical data concerning the density of eigenvalues. This is shown in Fig. 7 for the time-series of the different stocks of the S&P 500 (or other stock markets).

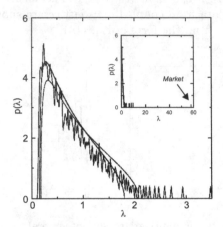

Fig. 7. Eigenvalue spectrum of the correlation matrices. Reprinted and adapted from L. Laloux, P. Cizeau, J.-P. Bouchaud, and M. Potters, *Noise Dressing of Financial Correlation Matrices*, Phys. Rev. Lett. **83**, 1467 (1999), Copyright (1997) by the American Physical Society.

Cross-correlations in financial data were also studied by Plerou et al. [23], who analyzed price fluctuations of different stocks through RMT. Using two large databases, they calculated cross-correlation matrices of returns constructed from: (i) 30-min returns of 1000 US stocks for the period 1994–95; (ii) 30-min returns of 881 US stocks for the period 1996–97; (iii) 1-day returns of 422 US stocks for the period 1962–96. They tested the statistics of the eigenvalues λ_i of cross-correlation matrices against a "null hypothesis" and found that a majority of the eigenvalues of the cross-correlation matrices were within the RMT bounds $(\lambda_{min}, \lambda_{max})$ defined above for random correlation matrices. Furthermore, they analyzed the eigenvalues of the cross-correlation matrices within the RMT bound for universal properties of random matrices and found good agreement with the results for the Gaussian orthogonal ensemble (GOE) of random matrices, implying a large degree of randomness in the measured cross-correlation coefficients. It was found that: (i) the distribution of eigenvector components, for the eigenvectors corresponding to the eigenvalues outside the RMT bound, displayed systematic deviations from the RMT prediction; (ii) such "deviating eigenvectors" were stable in time; (iii) the largest eigenvalue corresponded to an influence common to all stocks; (iv) the remaining deviating eigenvectors showed distinct groups, whose identities corresponded to conventionally-identified business sectors.

4 Approximate Entropy Method in Time-series Analysis

The Approximate Entropy (ApEn) method is an information theory-based estimate of the complexity of a time series introduced by S. Pincus [24], formally based on the evaluation of joint probabilities, in a way similar to the entropy of Eckmann and Ruelle. The original motivation and main feature, however, was not to characterize an underlying chaotic dynamics, rather to provide a robust model-independent measure of the randomness of a time series of real data, possibly – as it is usually in practical cases – from a limited data set affected by a superimposed noise. ApEn has been used by now to analyze data obtained from very different sources, such as digits of irrational and transcendental numbers, hormone levels, clinical cardiovascular time-series, anesthesia depth, EEG time-series, and respiration in various conditions.

Given a sequence of N numbers $\{u(j)\} = \{u(1), u(2), \ldots, u(N)\}$, with equally spaced times $t_{j+1} - t_j \equiv \Delta t = \text{const}$, one first extracts the sequences with embedding dimension m, i.e., $x(i) = \{u(i), u(i+1), \ldots, u(i+m-1)\}$, with $1 \le i \le N - m + 1$. The ApEn is then computed as

$$\text{ApEn} = \Phi^m(r) - \Phi^{m+1}(r), \tag{15}$$

where r is a real number representing a threshold distance between series, and the quantity $\Phi^m(r)$ is defined as

$$\Phi^m(r) = \langle \ln[C_i^m(r)] \rangle = \sum_{i=1}^{N-m+1} \ln[C_i^m(r)]/(N-m+1). \tag{16}$$

Here $C_i^m(r)$ is the probability that the series $x(i)$ is closer to a generic series $x(j)$ $(j \le N - m + 1)$ than the threshold r,

$$C_i^m(r) = \mathcal{N}[d(i,j) \le r]/(N - m + 1), \tag{17}$$

with $\mathcal{N}[d(i,j) \le r]$ the number of sequences $x(j)$ close to $x(i)$ less than r. As definition of distance between two sequences, the maximum difference (in modulus) between the respective elements is used,

$$d(i,j) = \max_{k=1,2,\ldots,m} (|u(j+k-1) - u(i+k-1)|). \tag{18}$$

Quoting Pincus and Kalman [25], "...ApEn measures the logarithmic frequency that runs of patterns that are close (within r) for m contiguous observations remain close (within the same tolerance width r) on the next incremental comparison". Comparisons are intended to be done at fixed m and r, the general ApEn(m,r) being in fact a family of parameters.

In economics, the ApEn method has been shown to be a reliable estimate of the efficiency of market [24–26] and has been applied to various economically relevant events. For instance, the ApEn computed for the S&P 500 index has shown a drastic increase in the two-week period preceding the stock market crash of 1987. Just before the Asian crisis of November 1997, the ApEn computed for the Hong Kong's Hang Seng index, from 1992 to 1998, assumes its highest values. More recently, a broader investigation carried out for various countries through the ApEn by Oh, Kim, and Eom, revealed a systematic difference between the efficiencies of the markets between the period before and after the the Asian crisis [27].

References

1. R.S. Tsay, Analysis of Financial Time Series, John Wiley, New York (2002). Econophysics and Sociophysics: Trends and Perspectives, B.K. Chakrabarti, A. Chakraborti, A. Chatterjee (Eds.), Wiley-VCH, Berlin (2006).
2. L. Bachelier, Théorie de la spéculation. Annales Scientifiques de l'Ecole Normale Supérieure, Suppl. 3, No. 1017, 21–86 (1900). English translation by A. J. Boness in: P. Cootner (Ed.), The Random Character of Stock Market Prices, Page 17, MIT, Cambridge, MA, (1967).
3. J.-P. Bouchaud, CHAOS **15**, 026104 (2005).
4. B.B. Mandelbrot, J. Business **36**, 394 (1963).
5. E. Fama, J. Business **38**, 34 (1965).
6. R. Cont, Quant. Fin. **1**, 223 (2001).
7. P. Gopikrishnan, M. Meyer, L.A.N. Amaral, H.E. Stanley, Eur. Phys. J. B **3**, 139 (1998).
8. R. Cont, M. Potters, J.-P. Bouchaud, in: B. Dubrulle, F. Graner and D. Sornette (Eds.), Scale Invariance and Beyond (Proc. CNRS Workshop on Scale Invariance, Les Houches, 1997), Springer, Berlin (1997).
9. Y. Liu, P. Cizeau, M. Meyer, C.-K. Peng, H.E. Stanley, Physica A **245**, 437 (1997); arxiv:cond-mat/9706021.
10. P. Cizeau, Y. Liu, M. Meyer, C.-K. Peng, H.E. Stanley, Physica A **245**, 441 (1997).
11. E.F. Fama, K.R. French, J. Fin. Economics **22**, 3 (1988).
12. R.N. Mantegna, H.E. Stanley, An Introduction to Econophysics, Cambridge University Press, New York (2000).
13. N. Vandewalle, M. Ausloos, Physica A **246**, 454 (1997). C.-K. Peng, S.V. Buldyrev, S. Havlin, M. Simons, H.E. Stanley, A.L. Goldberger, Phys. Rev. E **49**, 1685 (1994). Y. Liu, P. Gopikrishnan, P. Cizeau, M. Meyer, C.-K.. Peng, H.E. Stanley, Phys. Rev. E **60**, 1390 (1999). M. Beben, A. Orlowski, Eur. Phys. J. B **20**, 527 (2001). A. Sarkar, P. Barat, Physica A **364**, 362 (2006); arxiv:physics/0504038. P. Norouzzadeh, B. Rahmani, Physica A **367**, 328 (2006); D. Wilcox, T. Gebbie, arxiv:cond-mat/0404416.
14. See K. Kaneko (Ed.), Theory and Applications of Coupled Map Lattices, Wiley, New York (1993), and in particular the contribution of R. Kapral.
15. K. Kaneko, Physica D **34**, 1 (1989).
16. A. Chakraborti, M.S. Santhanam, Int. J. Mod. Phys. C **16**, 1733 (2005).
17. R.F. Engle, Econometrica **50**, 987 (1982).
18. T. Bollerslev, J. Econometrics **31**, 307 (1986).
19. P. Gopikrishnan, B. Rosenow, V. Plerou, H.E. Stanley, Phys. Rev. E **64**, 035106 (2001).

20. A.M. Sengupta, P.P. Mitra, Phys. Rev. E **60**, 3389 (1999).
21. V. Plerou, P. Gopikrishnan, B. Rosenow, L.A.N. Amaral, T. Guhr, H.E. Stanley, Phys. Rev. E **65**, 066126 (2002).
22. L. Laloux, P. Cizeau, J.-P. Bouchaud, M. Potters, Phys. Rev. Lett. **83**, 1467 (1999); `arxiv:cond-mat/9810255`.
23. V. Plerou, P. Gopikrishnan, B. Rosenow, L.A.N. Amaral, H.E. Stanley, Phys. Rev. Lett. **83**, 1471 (1999); `arxiv:cond-mat/9902283`. V. Plerou, P. Gopikrishnan, B. Rosenow, L.A.N. Amaral, T. Guhr, H.E. Stanley, Phys. Rev. E **65**, 066126 (2002); `arxiv:cond-mat/0108023`.
24. S. M. Pincus, Proc. Nati. Acad. Sci. USA **88**, 2297 (1991).
25. S. Pincus, R.E. Kalman, Proc. Nati. Acad. Sci. USA **101**, 13709 (2004).
26. S. Pincus, B.H. Singer, Proc. Nati. Acad. Sci. USA **93**, 2083 (1996).
27. G. Oh, S. Kim, C. Eom, Market efficiency in foreign exchange markets, Physica A, in press.

Correlations, Delays and Financial Time Series

K.B.K. Mayya and M.S. Santhanam

Physical Research Laboratory, Navrangpura, Ahmedabad 380 009, India

Summary. We study the returns of stock prices and show that in the context of data from Bombay stock exchange there are groups of stocks that remain moderately correlated for up to 3 days. We use the delay correlations to identify these groups of stocks. In contrast to the results of same-time correlation matrix analysis, the groups in this case do not appear to come from any industry segments. We present our results using the closing prices of 326 significant stocks of Bombay stock exchange for the period 1995 to 2005.

1 Introduction

The study of financial time series from the perspective of the models in physics is an active area of research. Much of the work is focussed on studying the time series from the stock markets and modelling them based primarily on the ideas of statistical physics. For instance, it is now broadly agreed that the distributions of returns from the stock markets are scale invariant [1]. In general, it is assumed that the time series of returns from the stocks are more or less like white noise. This implies that the successive return values are nearly uncorrelated. For instance, the auto-correlation of the returns would throw up a delta function at zero delay. In this article, we ask the question if the time series of returns can remain correlated with delays. Infact, it is known from several studies [2] that there are certain stocks which tend to display similar evolution. Often, they are stocks from similar industry segement, such as the automobile or technology companies. In this article, we study the correlation between the retruns of stock prices as a function of delays.

2 Correlated Stocks, Returns and Delays

In this work, we study the presence of delays in stock markets using the daily closing data from the Bombay Stock Exchange for the years 1995–2005. We start from the observation that certain stocks remain correlated for several years together. For instance, in Fig. 1, we show the stock prices of Infosys and Wipro which are strongly correlated over the entire period of the data under consideration. For any two standardised quantitites (*i.e*, mean zero and variance unity), $x(t)$ and $y(t)$, we define the cross-correlation as,

$$C(\tau) = \langle x(t)y(t+\tau)\rangle \tag{1}$$

where, τ is the delay and the angular bracket denotes the sample average. In the case of data shown in Fig. 1, the correlation at zero delay is $C(0) = 0.986$ and this provides a quantitative confirmation of the strong correlation between the two. However, notice that even though these stocks are strongly correlated we will not be able use this to any advantage since the returns of these stocks are not as strongly correlated. We define the returns for the stock x as,

$$r_x = \frac{x(t+1) - x(t)}{\sigma_x} \tag{2}$$

The returns for the stocks in Fig. 1 are shown in Fig. 2. We calculate the cross-correlation between them and the result is displayed in Fig. 3 . Clearly, the cross-correlation is significant only for values of $\tau \leq 3$ days. This implies that inspite of the delays of the order of 2–3 days, there exist some residual correlation among certain stocks. If the time series of returns were purely white noise, even a delay of one day would destroy all the correlations present in the data. The idea that returns contain some moderate correlations can be further confirmed by subjecting the data to detrended fluctuation analysis (DFA) [3]. The DFA is a technique to study the long range correlations present in non-stationary time series. For white noise, DFA gives an exponent of 0.5. The detrended fluctuation analysis performed on the returns for Wipro Ltd gives a slope of about 0.59 ± 0.01 which places it close to but not quite white noise exponent of 0.5. This preliminary analysis reveals that there could be groups of

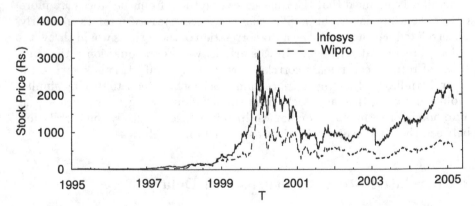

Fig. 1. The daily closing stock prices of Infosys and Wipro during 2001–2005. Note that they are strongly correlated.

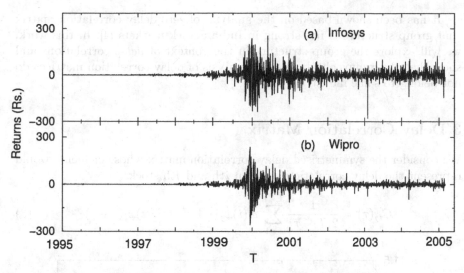

Fig. 2. The daily returns from the stocks of Infosys and Wipro during 2001–2005. As compared to Fig. 1, they are not as strongly correlated.

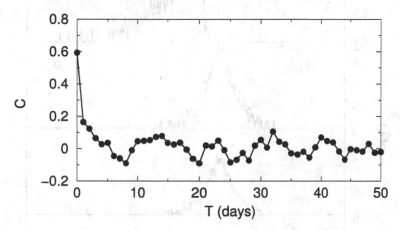

Fig. 3. The cross correlation between Infosys and Wipro as a function of time lag τ in days. Correlations continue to remain significant only for about 2–3 days.

stocks that could display residual cross correlation for short delay times. This provides the impetus to study the delay correlations among stocks. This can be elegantly done in the framework of delay correlation matrix. The study of delay correlation matrix and its random delay correlations have become important for several applications where lead-lag relationships are important. For instance, the U.S stock market was reported to show lead-lag relations between returns of large and small stocks [5]. have been few studies of random delay correlations in the last two years [6, 7].

It has been shown based on the analysis of zero delay correlation matrix that group structure is not strong in Indian stock markets [4]. In this work, we will explore the group structure in the context of *delay* correlations and show that the groups obtained from analysis of delay correlation matrices do not belong to same industry segments.

3 Delay Correlation Matrix

We consider the symmetrised delay correlation matrix whose elements would represent the delay correlation between ith and jth stock.

$$C_{ij}(\tau) = \frac{1}{2(T-\tau)} \sum_{t=1}^{T-\tau} x_i(t)x_j(t+\tau) + x_j(t)x_i(t+\tau) \tag{3}$$

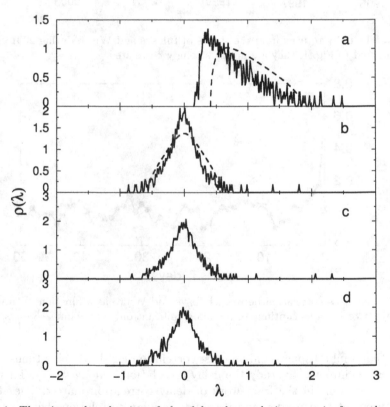

Fig. 4. The eigenvalue density of the delayed correlation matrix from the daily closing data of Bombay stock exchange. The dashed curve is the theoretical curve. (a) $\tau = 0$, (b) $\tau = 1$, (c) $\tau = 2$ and (d) $\tau = 3$. Note that three largest eigenvalues deviate strongly from the eigenvalue density for a random delay correlation matrix. As τ increases, the deviating eigenvalues are drawn closer to the bulk of the eigenvalue density.

This can be written in matrix notation as [6],

$$C(\tau) = \frac{M(0)M^{\mathrm{T}}(\tau) + M(\tau)M^{\mathrm{T}}(0)}{2T}. \tag{4}$$

where M is the data matrix of order $m \times n$. Assuming that the $C(\tau)$ is a random delay correlation matrix, Mayya and Amritkar [6] have obtained the analytical expression for the eigenvalue density for the case $m = n$ and it is given by, $\rho_\tau(\lambda) = 1/(2\pi\sigma^2)(\sqrt{2\sigma^2/|\lambda|} - 1)$, where σ is the standard deviation of the standardised data. In general, for $n > m$, $\rho_\tau(\lambda)$ will have to be numerically obtained by solving a fourth order algebraic equation. The case of delay correlation matrix without symmetrising has been considered by Biely and Thurner [7] and they have obtained the eigenvalue density in the general case of non-symmetric correlation matrix.

4 Dominant Modes and Delays

In Fig. 4 , we show the eigenvalue density for $\tau = 0, 1, 2, 3$ and the correponding "theoretical" curve obtained by simulating an ensemble of random delay correlation matrices. Notice that the largest three eigenvalues for $\tau = 1, 2, 3$

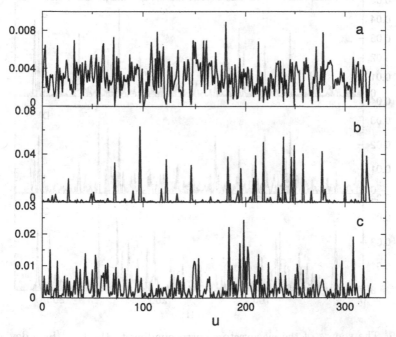

Fig. 5. The square of the eigenvectors corresponding to the first three deviating eigenvalues for $\tau = 1$ case. (a) λ_1, (b) λ_2, (c) λ_3.

also deviate significantly from the theoretical eigenvalue density shown as dashed line. The case of deviations of eigenvalues for $\tau = 0$ is already well documented in the literature [2]. The groups of stocks corresponding to these eigenvalues display significant correlations in the face of delays. Beyond, $\tau = 3$, no significant deviations are seen in the eigenvalue density. Hence, in the Indian stock market there are certain groups of stocks whose returns remain moderately correlated for about 2–3 days.

In order to identify the groups of stocks that contribute to the deviating eigenvalues, we study the eigenvectors corresponding to these eigenvalues. The largest eigenvalue λ_1 for all $\tau < 3$ corresponds to the market mode with almost all the stocks contributing to it. This is similar to the market mode identified in the case of correlation at zero delay. The next significant eigenvalues λ_2 and λ_3 at $\tau = 1$ and $\tau = 2$ contain signatures of the correlated groups of stocks. In this case, these group seem to remain correlated for delays of upto $\tau < 3$. However, these set of stocks that contribute significantly to λ_2 at $\tau = 1$ and $\tau = 2$ do not appear to belong to the same industry segment. The first three eigenvectors of the delay correlation matrix at $\tau = 1$ are shown in Fig. 5. Their eigenvalues deviate from RMT result as shown in Fig. 4(b). In Fig. 6, we show the eigenvectors of deviating eigenvalues for $\tau = 2$. Interestingly, the same set of stocks seem to contribute to all the deviating eigenvalues for $\tau = 1, 2$.

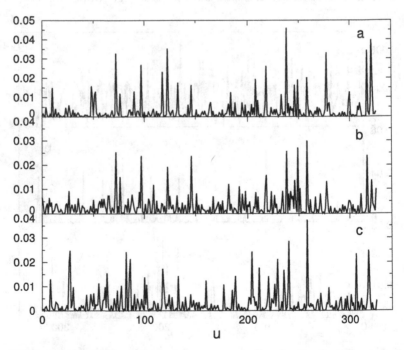

Fig. 6. The square of the eigenvectors corresponding to the first three deviating eigenvalues for $\tau = 2$ case. (a) λ_1, (b) λ_2, (c) λ_3.

5 Discussions

The time series of returns for certain groups of stocks show non-negligible correlations for short delays. In the case of daily closing data obtained from the Bombay stock exchange, the moderate correlations among the returns extends for about 2–3 days. The groups of stocks which show this tendency can be identified using the delay correlation matrix formalism. The results here would imply that modelling of returns should incorporate mechanisms for groups of stocks that display such mild correlations in empirical data. The emergence of group structure could be an important factor that differentiates developed and growing economies. In this context, it would be interesting to do a thorough analysis of delay correlation matrix with data from developed and emerging markets.

References

1. Gabaix X et. al. (2003) A Theory of Power-Law Distributions in Financial Market Fluctuations, Nature 423:267–270.
2. Plerou V et. al. (2002) Random Matrix approach to Cross-Correlations in Financial Data, Phys. Rev. E, 65:066126
3. Liu Y et. al. (1999) Statistical properties of the volatility of price fluctuations, Phys. Rev. E, 60:1390.
4. Sinha S, Pan RK (in this volume).
5. Lo AW, MacKinlay AC (1990) Review of Financial Studies 3:175–276.
6. Mayya KBK, Amritkar RE (2006) cond-mat/0601279.
7. Biely C, Thurner S (2006) physics/0609053.

Option Pricing with Log-stable Lévy Processes

Przemysław Repetowicz[1] and Peter Richmond[2]

[1] Probability Dynamics, IFSC House, Custom House Quay, Dublin 1, Ireland
[2] Department of Physics, Trinity College Dublin 2, Ireland. richmond@tcd.ie

Summary. We model the logarithm of the price (log-price) of a financial asset as
a random variable obtained by projecting an operator stable random vector with
a scaling index matrix $\underline{E}$ onto a non-random vector. The scaling index $\underline{E}$ models
prices of the individual financial assets (stocks, mutual funds, etc.). We find the
functional form of the characteristic function of real powers of the price returns
and we compute the expectation value of these real powers and we speculate on
the utility of these results for statistical inference. Finally we consider a portfolio
composed of an asset and an option on that asset. We derive the characteristic
function of the deviation of the portfolio, $\mathfrak{D}_t^{(t)}$, defined as a temporal change of
the portfolio diminished by the the compound interest earned. We derive pseudo-
differential equations for the option as a function of the log-stock-price and time and
we find exact closed-form solutions to that equation. These results were not known
before. Finally we discuss how our solutions correspond to other approximate results
known from literature,in particular to the well known Black & Scholes equation.

Key words: Option pricing, heavy tails, operator stable, fractional calculus

1 Introduction

Early statistical models of financial markets assumed that asset price returns
are independent, identically distributed (iid) Gaussian variables. [1]. However,
evidence has been found [2] that the returns exhibit power law (fat) tails in
the high end of the distribution. Except at very high frequencies or short times
([2]), a better statistical description for many financial assets is provided by
a model where the logarithm of the price is a heavy tailed one-dimensional
Lévy μ-stable process [3–6]. Since the tail parameter μ that measures the
probability of large price jumps will vary from one financial asset to the next,
a model based on operator stable Lévy processes [7] is appropriate. This model
allows the tail index to differ for each financial asset in the portfolio. Hence
we formulate a model where the log-price is a projections of an operator stable random vector onto a predefined direction (this projection determines the

portfolio mix). The cumulative probability distribution of the log-price diminishes as a mixture of power laws and thus the higher-order moments of the distribution may not exist and the characteristic function of the distribution may not be analytic.

Due to the constraints on the size of this paper we only include new results leaving proofs for further publications.

2 The Model of the Stock Market

In this section we define the model. In the following we recall certain known properties of operator stable distributions and we derive Fourier transforms of real powers of projections of operator stable vectors onto a non-random vector. In subsections (2.2) and (2.3) we derive Fourier transforms of operator stable random vectors for particular forms of parameters of the distribution.

2.1 The Basic Properties and New Results

Let $\log(S_t)$ be the logarithm of the price of the portfolio (log-price) at time t. We assume that the temporal change of the log-price is composed of two terms, a deterministic term and a fluctuation term viz:

$$\frac{dS_t}{S_t} = \frac{S_{t+dt} - S_t}{S_t} = \alpha dt + \boldsymbol{\sigma} \cdot d\mathbf{L}_t \qquad (1)$$

The parameters $\alpha \in \mathbb{R}$ (the drift) and the elements of the D dimensional vector $\boldsymbol{\sigma} := (\sigma_1, \ldots, \sigma_D)$ (the portfolio mix) are assumed to be non-random constants. The random vector $\mathbf{L}_t$ is (strictly) operator stable, meaning that it is an operator-normalized limit of a sum of some independent, identically distributed (iid) random vectors $\mathbf{X}_i$. We have

$$\mathbf{L}_t := \lim_{n \to \infty} n^{-E} \sum_{i=1}^{\lfloor nt \rfloor} \mathbf{X}_i \qquad (2)$$

where $n^{-E} = \exp(-E \ln n)$ and E is a real D-dimensional matrix such that the equality holds in distribution. The class of distributions of the former vectors related to a given matrix E is termed an attraction domain of an operator stable law. Members of such class are usually unknown.

We now recall some known facts [7] concerning the operator stable probability density $\omega_{\mathbf{L}_t}$ and its Fourier transform $\tilde{\omega}(\mathbf{k}) := \mathcal{F}_\mathbf{x}[\omega](\mathbf{k})$.

The following identities hold:

$$\omega_{\mathbf{L}_t}(\mathbf{x}) = \omega_{\mathbf{L}_1}\left(t^{-E}\mathbf{x}\right) \det\left(t^{-E}\right) \quad \text{for all } t > 0. \qquad (3)$$

and

$$\tilde{\omega}^t(\mathbf{k}) = \exp\left(-t\phi(\mathbf{k})\right) = \tilde{\omega}\left(t^{E^T}\mathbf{k}\right) = \tilde{\omega}_{\mathbf{L}_t}(\mathbf{k}) \tag{4}$$

where E^T is the transpose of E and ϕ is the negative logarithmic characteristic function of the random vector $\mathbf{L}_1$. In the following we assume that that function is even:

$$\phi(\mathbf{k}) = \phi(-\mathbf{k}) \tag{5}$$

The identities (3) are termed as a self-similar property of the random walk $\mathbf{L}_t$.

A motivation for introducing model (1) is statistical inference of parameters of a distribution that describes real financial data. In this context it is useful to know analytically the distribution of a real power of the integrated fluctuation term in (1). In general this is not known. Here we derive some new results for operator stable Lévy distributions. Denote by $\nu_t^{(\beta)}(z)$ the pdf of a random process $\Xi_t^{(\beta)} : (\boldsymbol{\sigma} \cdot \mathbf{L}_t)^\beta$ and by $\tilde{\nu}_t^{(\beta)}(k)$ its Fourier transform. For $\beta \geq 1$ the identity holds:

$$\tilde{\nu}_t^{(\beta)}(k) = \begin{cases} \displaystyle\int_{-\infty}^{\infty} d\lambda\, \tilde{\omega}\left(t^{E^T}\hat{\sigma}\lambda\right) \mathcal{K}^{(\beta)}(\mathfrak{k},\lambda) & \text{for } \beta \neq 1 \\[2ex] \tilde{\omega}\left(t^{E^T}\boldsymbol{\sigma}k\right) & \text{for } \beta = 1 \end{cases} \tag{6}$$

where

$$\mathcal{K}^{(\beta)}(k,\lambda) := \begin{cases} \displaystyle\frac{1}{2\pi}\int_0^{\infty} dt\, e^{-kt^\beta}\left(\mathfrak{r}e^{-\imath\lambda\mathfrak{r}t} + \mathfrak{r}\mu e^{\imath\lambda\mathfrak{r}\mu t}\right) & \text{if } \lfloor\beta\rfloor \text{ is even} \\[3ex] \displaystyle\frac{1}{2\pi}\int_0^{\infty} dt\, e^{-kt^\beta}\left(\mathfrak{r}e^{-\imath\lambda\mathfrak{r}t} + \bar{\mathfrak{r}}\mu e^{\imath\lambda\bar{\mathfrak{r}}\mu t}\right) & \text{if } \lfloor\beta\rfloor \text{ is odd} \end{cases} \tag{7}$$

and $\hat{\sigma} := \boldsymbol{\sigma}/|\boldsymbol{\sigma}|$ and $\mathfrak{r} := \exp(\imath\pi/(2\beta))$, $\mu := \exp(-\imath\pi\{\beta\}/(\beta))$. The symbols $\lfloor\beta\rfloor$ and $\{\beta\}$ mean the biggest integer not larger then β and the fractional part of β respectively, and $\mathfrak{k} := k\sigma^\beta$.

In addition for even values of $\lfloor\beta\rfloor$ we have:

$$\tilde{\nu}_t^{(\beta)}(k) = \frac{1}{\pi}\int_{l(\mathfrak{r},\beta)} dz\, e^{-z}\int_0^{\infty} d\xi \left(\frac{\sin(\xi)}{\xi}\right) \tilde{\omega}\left(t^{E^T}\left(\frac{k}{z}\right)^{1/\beta}\frac{\xi}{\mathfrak{r}}\boldsymbol{\sigma}\right) \tag{8}$$

where the integration line $l(\mathfrak{r},\beta))$ reads:

$$l(\mathfrak{r},\beta) := \left[0, \mathfrak{r}^{-\beta}\infty\right] \cup \left[0, (-1)^\beta \mathfrak{r}^{-\beta}\infty\right] \tag{9}$$

The identities (8) and (9) may be useful for describing the magnitude of the fluctuations of a random walk.

Thus it follows that the fractional moments of the scalar product are obtained by differentiating the Fourier transform at $k = 0$. We will obtain closed form results for these moments in section (2.4). Here we only recall that in the non-Gaussian case, due to (5), we have:

$$E\left[(\boldsymbol{\sigma} \cdot \mathbf{L_t})^\beta\right] = \left\{ \begin{array}{l} \infty \text{ when } \beta \text{ is even} \\ 0 \text{ when } \beta \text{ is odd} \end{array} \right\} \tag{10}$$

In addition for $\beta > 0$ there the moment exists only if β does not exceed a certain threshold value.

It is our objective to price options on the portfolio of stocks driven by operator stable fluctuations (see section (3)). By this we mean a theory that 1) allows inference of the the stable index E and the Lévy measure of the whole vector of stock prices (market)from a statistical sample and 2)hedges against risk in the market by the construction of an appropriate option. To the best of our knowledge, this has not yet been achieved.

The generic properties of operator stable probability distributions and their marginals are described in [7]. Here we recall some known facts and we analyse two particular cases of the stable index. The Fourier transform $\tilde{\omega}(\mathbf{k})$ is uniquely determined via the stable index E and the log-characteristic function ϕ confined to a unit sphere. This can be seen by representing the vector $\mathbf{k}$ in the Jurek coordinates viz

$$\mathbf{k} = r_\mathbf{k}^{E^T} \cdot \boldsymbol{\theta_k} \tag{11}$$

where $|\boldsymbol{\theta_k}| = 1$. Using the scaling relation (4) we get:

$$\exp\left(-r_\mathbf{k}\phi\left(\boldsymbol{\theta_k}\right)\right) = \tilde{\omega}^{r_\mathbf{k}}\left(\boldsymbol{\theta_k}\right) = \tilde{\omega}\left(r_\mathbf{k}^{E^{(T)}} \cdot \boldsymbol{\theta_k}\right) = \tilde{\omega}(\mathbf{k})\exp(-\phi(\mathbf{k})) \tag{12}$$

and thus

$$\phi(\mathbf{k}) = r_\mathbf{k}\phi\left(\boldsymbol{\theta_k}\right) \tag{13}$$

Since it follows from the Jordan decomposition theorem that every matrix E is, in a certain basis, a block diagonal matrix the set of all possible jump intensities ω is narrowed down to few classes of solutions only, each one corresponding to a particular Jordan decomposition of the matrix E. We now firstly investigate a few classes of solutions as a function of E and subsequently the generic solution for an arbitrary E. The existence results in this field are given in [7]. We stress that, contrary to [7], we aim at computing the characteristic functions and the fractional moments in closed form rather than only showing their existence.

2.2 Pure Scaling

In this case $E = (D\mu)^{-1}I$ where I is a D dimensional identity matrix and $\mu > 0$ is a constant. From (11) we see that:

$$\mathbf{k} = r_\mathbf{k}^{1/(D\mu)} \theta_\mathbf{k} = |\mathbf{k}| \frac{\mathbf{k}}{|\mathbf{k}|} \tag{14}$$

hence $r_\mathbf{k} = |\mathbf{k}|^{D\mu}$ and $\theta = \mathbf{k}/|\mathbf{k}|$ and so

$$\tilde{w}(\mathbf{k}) = \exp\left(-|\mathbf{k}|^{D\mu} \phi\left(\frac{\mathbf{k}}{|\mathbf{k}|}\right)\right) \tag{15}$$

The β-marginal probability density function from (6) reads:

$$\tilde{\nu}_t^{(\beta)}(k) = \int_{-\infty}^{\infty} d\lambda \exp\left\{-t|\lambda|^{D\mu} \phi\left(\frac{\sigma}{|\sigma|}\text{sign}(\lambda)\right)\right\} \mathcal{K}^{(\beta)}(k\sigma^\beta, \lambda) \tag{16}$$

where the kernel $\mathcal{K}$ is defined in (7). From the properties of the Gamma function we obtain easily the fractional moment of the scalar product as:

$$E\left[(\sigma \cdot \mathbf{L}_t)^\beta\right] = \mathfrak{C}(\beta, D\mu) \cdot \left(\sigma t^{(D\mu)^{-1}}\right)^\beta \cdot \left(\cos\left(\frac{\pi\beta}{2}\right) e^{i\frac{\beta\pi}{2}}\right) \cdot \phi_\pm^{\beta/(D\mu)} \tag{17}$$

where

$$\mathfrak{C}(\beta, D\mu) := \left(\frac{2^{\beta+1}}{D\mu\sqrt{\pi}} \Gamma\left(\frac{\beta+1}{2}\right) \frac{\Gamma\left(-\frac{\beta}{D\mu}\right)}{\Gamma\left(-\frac{\beta}{2}\right)}\right) \tag{18}$$

Here $\phi_\pm := \phi(\pm)$. The moment exists for $\beta < D\mu$. The prefactor (18) in (17) fits in with the known result for the fractional moment of a modulus of a stable variable (see equation (3.6) page 32 in [9]). For the derivations of that result by means of the Mellin-Stieljes transform see [25, 26] and by means of characteristic functions see [27].

2.3 Scaling & Rotation

In this case $D = 2$ and we chose:

$$E = \begin{pmatrix} (2\mu)^{-1} & -b \\ b & (2\mu)^{-1} \end{pmatrix} \tag{19}$$

Clearly the trace $\text{Tr}[E] = \mu^{-1}$. We denote by $O_\beta := \begin{pmatrix} \cos(\beta) & -\sin(\beta) \\ \sin(\beta) & \cos(\beta) \end{pmatrix}$ a two dimensional rotation by an angle $\beta \in \mathbb{R}$. The mapping:

$$r^{E^T} : \mathbb{R}^2 \ni \mathbf{k} \to r^{(2\mu)^{-1}} O_{-b\log(r)}\mathbf{k} \in \mathbb{R}^2 \tag{20}$$

changes the length of $\mathbf{k}$ by a multiplicative factor $r^{(2\mu)^{-1}}$ and rotates the vector by an angle $-b\log(r)$. The Jurek coordinates read: $r_\mathbf{k} = |\mathbf{k}|^{2\mu}$ and $\theta_\mathbf{k} = O_{2b\mu\log(|k|)}(\mathbf{k}/|\mathbf{k}|)$ and so

$$\tilde{w}(\mathbf{k}) = \exp\left(-|\mathbf{k}|^{2\mu} \phi\left(O_{2b\mu\log(|k|)}\frac{\mathbf{k}}{|\mathbf{k}|}\right)\right) \tag{21}$$

The β-marginal probability density function and the fractional moment read:

$$\tilde{\nu}_t^{(\beta)}(k) = \int_{-\infty}^{\infty} d\lambda \exp\left\{ -t|\lambda|^{2\mu} \phi\left(O_{2b\mu \log(|\lambda|)} \frac{\sigma}{|\sigma|} \right) \right\} \mathcal{K}^{(\beta)}(k\sigma^\beta, \lambda) \quad (22)$$

and

$$E\left[(\sigma \cdot \mathbf{L}_t)^\beta \right] = \mathfrak{C}(\beta, 2\mu) \cdot \left(\sigma t^{(2\mu)^{-1}} \right)^\beta \cdot \left(\cos\left(\frac{\pi\beta}{2} \right) e^{i\frac{\beta\pi}{2}} \right)$$

$$\cdot \frac{1}{2\pi} \int_0^{2\pi} d\eta \phi^{\beta/(2\mu)}(O_\eta \hat{\sigma}). \quad (23)$$

respectively. Here $\mathfrak{C}(\beta, 2\mu)$ is defined in (18). The moment exists for $\beta < 2\mu$. Since the moment depends on the average of a power of the log-characteristic function over the unit sphere we conclude that the knowledge of the moments does not determine the distribution in a unique manner.

In the following section we compute the fractional moments of the scalar product in the generic case of a operator stable distribution.

2.4 The Generic Case

Assume that the stable index has D different eigenvalues $\{\lambda_p\}_{p=1}^D$ that are either real or pairwise complex conjugate. Then the following spectral decomposition holds:

$$\underline{E}^T \underline{Q} \cdot \mathrm{Diag}(\lambda) = \cdot \underline{Q}^{-1} \quad (24)$$

where $\mathrm{Diag}(\lambda) := (\delta_{i,j}\lambda_j)_{i,j=1}^D$ and such that the matrix $\underline{Q}$ is unitary

$$\underline{Q} \cdot \underline{Q}^\dagger = \underline{Q}^\dagger \cdot \underline{Q} = 1 \quad (25)$$

Then from (24), from the definition of the operator $r^{\underline{E}^T} := \exp(\underline{E}^T \log(r))$ and from the Cayley-Hamilton theorem we easily arrive at the identity:

$$r^{\underline{E}^T} := \underline{Q} \cdot \mathrm{Diag}(r^\lambda) \cdot \underline{Q}^{-1} \quad (26)$$

where $\mathrm{Diag}(r^\lambda) := (\delta_{i,j} r^{\lambda_j})_{i,j=1}^D$ From (11) and (26) we obtain following equations for the Jurek coordinates $r := r_{\hat{\sigma}\lambda}$ and $\hat{\theta}_r := \theta_{\hat{\sigma}\lambda}$ of the vector $\hat{\sigma}\lambda$. We denote $\lambda_j := \Theta_j + i\Upsilon_j$ and we have:

$$\lambda = \left(\sum_{j=1}^D \frac{|\tilde{\sigma}_j|^2}{r^{2\Theta_j}} \right)^{-\frac{1}{2}} \left| r^{-\underline{E}^T} \hat{\sigma} \right|^{-1} \quad \text{and} \quad \hat{\theta}_r = \lambda \left(\sum_{j=1}^D \underline{Q}_{i,j} \frac{\tilde{\sigma}_j}{r^{\lambda_j}} \right)_{i=1}^D \lambda r^{-\underline{E}^T} \hat{\sigma}$$

$$(27)$$

where $\tilde{\sigma}_i := \underline{Q}_{i,j}^{-1}\hat{\sigma}_j$ are projections of the unit vector $\hat{\sigma}$ onto the eigenvectors of the stable index (rows of the matrix $\underline{Q}^{-1}$ or columns of the matrix $\underline{Q}$). If the unit vector is proportional to the lth eigenvector then $\tilde{\sigma}_i = \delta_{i,l}$ and from (27) we get $r = \lambda^{\Theta_l^{-1}}$ and $\hat{\boldsymbol{\theta}}_r = \lambda^{-\imath \Upsilon_l/\Theta_l}\hat{\sigma}$.

The fractional moment reads:

$$E\left[(\boldsymbol{\sigma} \cdot \mathbf{L}_t)^\beta\right] = \mathfrak{C}\left(\beta, \Theta_l^{-1}\right) \cdot \left(|\boldsymbol{\sigma}| \, t^{\Theta_l}\right)^\beta \cdot \left(\cos\left(\frac{\pi}{2}\beta\right) e^{\imath \frac{\pi}{2}\beta}\right)$$

$$\times \frac{1}{2\pi} \int_0^{2\pi} d\xi \phi^{\beta \Theta_l}\left(e^{-\imath \xi}\hat{\sigma}\right) 1_{\sum_{j \neq J} \tilde{\sigma}_j^2 |\phi(e^{-\imath \xi}\hat{\sigma})|^{2\Theta_j} \leq |\sigma|^2 |\phi(e^{-\imath \xi}\hat{\sigma})|^{2\Theta_l}} \quad (28)$$

Here l is such a number that $\Theta_l - \Theta_j \geq 0$ for all $j \neq l$ such that $|\tilde{\sigma}_j| > 0$, and $J := \{j \,|\, \Theta_j = \Theta_l\}$, $D^* = D - \text{card}(J) + 1$, and $\mathfrak{C}(\beta, \Theta_l^{-1})$ is defined in (18). The moment exists for $\beta \Theta_l < 1$, meaning if β does not exceed the inverse biggest real part of eigenvalues of the tail index $\underline{E}$ in the maximal eigenspace containing $\hat{\sigma}$. We note the following:

[1] The moment is a product of a real prefactor, the βth power of the length of the vector $\boldsymbol{\sigma}$, a time factor $t^{\beta \Theta_l}$ and a complex prefactor. The former prefactor is the same as in the one dimensional case whereas the later prefactor is a complex number equal to the support of the random variable $(\boldsymbol{\sigma} \cdot \mathbf{L}_t)^\beta$. In particular for symmetric distributions the later factor is real and the support of random variable is given by $1 + (-1)^\beta = \cos(\frac{\pi}{2}\beta)e^{\imath \frac{\pi}{2}\beta}$.

[1] The integrand in (28) is related to the complement of the J-space, meaning a linear span of eigenvectors whose real parts of eigenvalues equal Θ_l.

[2] The result (28) is in accordance with an existence result (Theorem 8.3.10 in [7]). However we have for the first time computed the moment in closed form which will be useful for statistical inference for example or for other theoretical work.

[3] If the multiplicity of Θ_l is equal to D then $\text{card}(J) = D$ and the complement of J is empty and the last term in (28) reduces to

$$\langle \phi^{\beta \Theta_l} \rangle := (2\pi)^{-1} \int_0^{2\pi} d\xi \phi^{\beta \Theta_l}\left(e^{-\imath \xi}\hat{\sigma}\right) \quad (29)$$

because the Heaviside function in the integrand is identically equal unity. This is like in the scaling & rotation case.

[4] If $\phi(e^{-\imath \xi_l}\hat{\sigma}) < 1$ then the left hand side of the equality in the subscript of the Heaviside function is positive and the Heaviside function may not be identically equal unity.

3 The Option Price

An option on a financial asset is an agreement settled at time t to purchase (call) or to sell (put) the asset at some maturity time T in the future. Here we consider European style options that can be exercised only at maturity. This means that boundary conditions are imposed on the option price at maturity $t = T$. Extending the analysis to American style options that may be exercised at any time can be done by considering European style options with a different number of exercise times [10] and allowing the number of exercise times to go to infinity.

In order to minimize the risk we now divide the money available between N_S stocks S_t and N_C options $C(S_t; t)$. The value of the portfolio is then:

$$V(t) = N_S S_t + N_C C(S_t; t) \tag{30}$$

We may, without loss of generality, chose $N_C = 1$.

The portfolio is a stochastic process that is required to grow exponentially with time in terms of its expectation value. The rate of growth r is the so-called 'riskless' rate of interest and is assumed to be independent of time t.

3.1 Local Temporal Growth

Consider the distribution of deviations

$$\mathfrak{D}_t^{(dt)} := V(t + dt) - e^{rdt}V(t) \tag{31}$$

between the interest $(e^{rdt} - 1)V(t) = (rdt + O(dt^2))V(t)$ that is earned by the portfolio and the change $V(t + dt) - V(t)$ of the price of the portfolio. Does a self-financing strategy exists? Is it possible to choose $C = C(S_t; t)$ subject to a condition $C_T = \max(S_T - K, 0)$, for some strike price K, such that the expectation value of the deviations of the portfolio conditioned on the price of the stock at time t equals zero? Thus we require that the deviations have no drift:

$$E\left[\mathfrak{D}_t^{(dt)} | S_t\right] = 0 \tag{32}$$

In our model we assume that the above condition is satisfied only for an infinitesimal time change dt and is conditioned on the value of the stock price at time t (local temporal growth sec. 3.1). Due to limited space we are not able to include a model extension that assumes that the above condition is satisfied for a finite dt We will present it in a future publication.

Our approach is more general than that used in financial mathematics [11] where considerations are based on the lack of arbitrage, meaning the assumption that riskless opportunities for making money in financial transactions do not exist. We waive that unrealistic assumption and instead require the portfolio to increase exponentially with time.

From equation (1) we have:

$$S_{t+dt} - S_t = S_t \left(\exp\left[\alpha dt + \boldsymbol{\sigma} \cdot \mathbf{L}_{dt}\right] - 1\right) \tag{33}$$

where we have used the fact that a Lévy process is homogeneous in time, meaning that

$$\mathbf{L}_{t+dt} - \mathbf{L}_t \stackrel{d}{=} \mathbf{L}_{dt} \tag{34}$$

where $\stackrel{d}{=}$ in (34) means an equality is in distribution. We note that (33) is merely a transformation of equation (1) and not a solution to that equation. As such equation (33) holds for infinitesimal times dt only.

From (33) we see that the expectation value of the right hand side conditioned on S_t and is infinite, unless the fluctuations are Gaussian.

Therefore we will modify the log-characteristic function $\phi(\boldsymbol{\lambda})$ in order to ensure the finiteness of all moments. We define:

$$\phi_\epsilon(\boldsymbol{\lambda}) := \phi(\boldsymbol{\lambda}) \exp\left(-\frac{\epsilon}{|\boldsymbol{\lambda}|}\right) \tag{35}$$

where $\epsilon > 0$ and replace ϕ_ϵ by ϕ. From now on we will work with a fictious process related to the modified log-characteristic function, we will solve the option pricing problem for it and at the end of the derivation we will take the limit $\epsilon \to 0$. After finishing the derivation we will check analytically if the result ensures a risk free portfolio. Firstly we check that conditional expectation value of the stock price is finite. We have:

$$E\left[S_{t+dt} - S_t \mid S_t\right] = S_t \left(\exp\left(\alpha dt\right) E\left[e^{(\boldsymbol{\sigma} \cdot \mathbf{L}_{dt})}\right] - 1\right)$$
$$= S_t \left(\exp\left(\alpha dt\right) e^{-dt\phi(-\imath\boldsymbol{\sigma})} - 1\right) \leq \infty \tag{36}$$

where in the second equality in (36) we used the following identity:

$$E\left[e^{(\boldsymbol{\sigma} \cdot \mathbf{L}_{dt})}\right] = \int_{\mathbb{R}} dz e^z \delta\left(z - \xi\right) \nu_{\boldsymbol{\sigma} \cdot \mathbf{L}_{dt}}(\xi) d\xi$$

$$= \int_{\mathbb{R}} dz e^z \frac{1}{2\pi} \int_{\mathbb{R}} dk e^{\imath k(\xi - z)} \cdot \nu_{\boldsymbol{\sigma} \cdot \mathbf{L}_{dt}}(\xi) d\xi \tag{37}$$

$$= \int_{\mathbb{R}} dk \delta(k + \imath) \tilde{\nu}_{\boldsymbol{\sigma} \cdot \mathbf{L}_{dt}}(k)$$

$$= \int_{-\imath + \mathbb{R}} dk \delta(k + \imath) \tilde{\nu}_{\boldsymbol{\sigma} \cdot \mathbf{L}_{dt}}(k) e^{-dt\phi(-\imath\boldsymbol{\sigma})} \tag{38}$$

In the first equality in (37) we inserted a delta function into the definition of the expectation value, in the second equality we used the integral representation of the delta function, in the first equality in (38) we integrated over z and ξ and we used the integral representation of the delta function in the second equality we shifted the integration line by using the Cauchy theorem applied to a rectangle $[-R, R] \cup R + \imath[0, 1] \cup -\imath + [R, R] \cup R - \imath[0, 1]$ in the limit $R \to \infty$ and in the last equality we used (6) and (4). We make three remarks. Firstly the delta function has been analytically continued to complex arguments, ie we have defined it as follows:

$$\delta(k + \imath q) := \sum_{p=0}^{\infty} \frac{(\imath q)^p}{p!} \delta^{(p)}(k) \tag{39}$$

Secondly we note that the result (38) holds only for $\epsilon > 0$ because otherwise, all $p \geq 2$ terms in the sum (39) produce infinite values when integrated with the second term in the integrand. Thirdly we reiterate that it is the fictious, modified stock price, related to $\epsilon > 0$, that has a finite expectation value whereas the real stock price has of course an infinite expectation value. The option prices that we compute correspond to a fictious ϵ-world where stock prices' probabilities have been modified like in (35) and the real option price is obtained as a limit of the former as ϵ tends to zero and the sequence of the fictious worlds towards our real world. In other words the option pricing problem has indeed no solution in our real world however it has a solution in the "complement" of our world by the limit of the ϵ-worlds.

We will therefore construct a zero-expectation value stochastic process (31) as a linear combination (30) of two stochastic processes S_t and $C(S_t; t)$ that have both non-zero expectations values. For this purpose we will analyze the probability distribution of the deviation variable $\mathfrak{D}_t^{(dt)}$ and work out conditions for the option price such that the conditional expectation value $E\left[\mathfrak{D}_t^{(dt)} | S_t\right]$ is equal zero.

$$\mathfrak{D}_t^{(dt)} = (V(t+dt) - V(t)) + V(t)\left(1 - e^{rdt}\right) \tag{40}$$

$$= N_S\left(S_{t+dt} - S_t\right) + (C_{t+dt} - C_t) + V(t)\left(1 - e^{rdt}\right) \tag{41}$$

$$= N_S S_t\left(e^{\alpha dt + \sigma \cdot \mathbf{L}_{dt}} - 1\right) + (C(S_{t+dt}; t+dt) - C_t) + V(t)\left(1 - e^{rdt}\right) \tag{42}$$

$$= N_S S_t\left(e^{\alpha dt + \sigma \cdot \mathbf{L}_{dt}} - 1\right) + \frac{\partial C}{\partial t} dt + \sum_{m=1}^{\infty} \frac{1}{m!} \frac{\partial^m C}{\partial S^m} S_t^m \left(e^{\alpha dt + \sigma \cdot \mathbf{L}_{dt}} - 1\right)^m$$

$$+ V(t)\left(1 - e^{rdt}\right) \tag{43}$$

In (43) we have expanded the price of the option in a Taylor series to the first order in time and to all orders in the price of the option. In that we have assumed that the price of the option is a perfectly smooth function of the price of the stock. This may limit the class of solutions. In particular, solutions may exist, where the price of the stock is a function satisfying the Hölder condition:

$$|C(S_{t+dt}; t+dt) - C_t| \leq A |S_{t+dt} - S_t|^{\Lambda} \tag{44}$$

for any S_{t+dt} and S_t, a constant A and a Hölder exponent $\Lambda \in [0, 1)$ and thus the price of the option can be expanded in a fractional Taylor series [14] in powers of $S_{t+dt} - S_t$. We will seek for these solutions in future work.

The process $\mathfrak{D}_t^{(dt)}$ is a sum of infinitely many terms that have non-zero expectation values. We could compute its expectation value directly using (36) and re-sum the series. However, we will instead calculate the characteristic function of the process $\mathfrak{D}_t^{(dt)}$ conditioned on the value of the process S_t at time t. This means that we propagate the process S_t by an infinitesimal value dt and we compute the characteristic function of the increment and we require the zero value derivative of the characteristic function to be equal zero. This technique is not new, see discussion about solving master equations of Markov processes in [17], and it works because of the time-homogeneity of the process (34) and because of the fact that the parameters σ and α are constant as a function of the process S_t. The time-homogeneity follows from the infinite divisibility of the process and thus the technique applied here also works in the generic setting of Lévy processes. We will extend the model according to these lines in future investigations.

We note that $\mathfrak{D}_t^{(dt)}$ in (43) is a function of the scalar product $\sigma \cdot \mathbf{L}_{dt}$ only and thus the distribution of $\mathfrak{D}_t^{(dt)}$ is unique functional of the distribution of the scalar product.

We derive the distribution of $\mathfrak{D}_t^{(dt)}$ now. We define

$$D_t := \left(\partial_t C dt + V(t) \left(1 - e^{rdt} \right) \right) \tag{45}$$

we condition on the value of the fluctuation $\sigma \cdot \mathbf{L}_{dt}$, we use (43), and we get:

$$\chi_{\mathfrak{D}_t^{(dt)}|S_t}(k) = \exp\left(\imath k D_t \right)$$

$$\times \left(1 + \sum_{m=1}^{\infty} \frac{(\imath k)^m}{m!} \sum_{q=0}^{\infty} \mathfrak{A}_m(q) \exp\left(-dt \left(\phi\left(-\imath q \sigma \right) - q\alpha \right) \right) \right) \tag{46}$$

where ϕ is the negative logarithmic characteristic function of the random vector $\mathbf{L}_1$ (see (4)). The expectation value of the portfolio deviation conditioned on the value of the price of the stock S_t reads:

$$E\left[\mathfrak{D}_t^{(dt)} | S_t \right] = \left. \frac{d\chi_{\mathfrak{D}_t^{(dt)}|S_t}(k)}{d(\imath k)} \right|_{k=0} \tag{47}$$

$$= \left(\partial_t C - rV(t) - \left(N_S + \frac{\partial C}{\partial S} \right) S_t \left(\mathfrak{S}_{1,\phi} - \alpha \right) - \sum_{n=1}^{\infty} \mathfrak{E}_n \frac{\partial^n C}{\partial (\log(S))^n} \right) dt$$

$$+ O\left(dt^2 \right) \tag{48}$$

where

$$\mathfrak{S}_{n,\phi} := \sum_{q=0}^{n} \binom{n}{q} (-1)^{n-q} \phi(-\imath q \sigma) \quad \text{and} \quad \mathfrak{E}_n = \sum_{k=\max(n,2)}^{\infty} \frac{a_n^{(k)}}{k!} \mathfrak{S}_{k,\phi} \tag{49}$$

and the log-characteristic function $\phi(-\imath q \sigma)$ in (48) has been analytically continued to imaginary arguments. Here the coefficients $a_n^{(k)}$ read:

$$a_n^{(k)} := (-1)^{k-n}(k-1)! \sum_{1 \le j_1 < \cdots < j_{n-1} \le k-1} \prod_{q=1}^{n-1} \frac{1}{j_q}$$

$$= (-1)^{k-n} \sum_{1 \le j_1 < \cdots < j_{k-n} \le k-1} \prod_{q=1}^{k-n} j_q \tag{50}$$

with $a_1^{(k)} = (-1)^{k-1}(k-1)!$. In addition the coefficients $\mathfrak{E}_n$ satisfy:

$$\sum_{n=1}^{\infty} \mathfrak{E}_n = 0 \tag{51}$$

what follows readily from the fact that $\sum_{n=1}^{k} a_n^{(k)} = 0$ for $k \ge 2$.

In the Gaussian case the coefficients read $\mathfrak{S}_{n,\phi} = (-\sigma)^2 (n\delta_{n,1} + n(n-1) \times \delta_{n,2})$ and thus (48) yields a second order PDE. Since the Levy distribution has been truncated as in (35) and due to (5) the result in (49) is real. Indeed the log-characteristic function can be expanded in a Taylor series in even powers of the argument only and thus its value at the negative imaginary unit is real. If we did not truncate we would have obtained a unrealistic complex result as seen from (15). We reiterate that the limit of truncation threshold going to zero ($\epsilon \to 0$ in (35)) will be taken at the end of the calculation only rather than at intermediary stages. If we did so at this stage we would have obtained a paradoxical result; an infinite sum of numbers $\mathfrak{E}_n$ each of which is infinite equals zero.

The requirement $E\left[\mathfrak{D}_t^{(dt)} | S_t\right] = O(dt)$ implies a following generalized Black & Scholes equation:

$$\partial_t C - \sum_{n=2}^{\infty} \frac{\mathfrak{S}_{n,\phi}}{n!} (S_t)^n \frac{\partial^n C}{\partial S^n} = \partial_t C - \sum_{n=1}^{\infty} \mathfrak{E}_n \frac{\partial^n C}{\partial (\log(S))^n}$$

$$= rV(t) + (N_S + \frac{\partial C}{\partial S}) S_t (\mathfrak{S}_{1,\phi} - \alpha) \tag{52}$$

In order that we get further insight into the problem, in particular in order that we are able to solve equation (52) analytically we find a new expression for the coefficients of the PDE stated in the following propositions.

Proposition 1. *The coefficients $\mathfrak{S}_{k,\phi}$ in (52) read:*

$$\mathfrak{S}_{k,\phi} = \int_0^{\infty} d\xi \tilde{\phi}(\xi) \left(-1 + e^{-\xi}\right)^k \tag{53}$$

for $m \in \mathbb{N}$. Here $\tilde{\phi}(\xi)$ is the inverse Laplace transform of the log-characteristic function of $\mathbf{L}_1$ or the Lévy measure of the process $\mathbf{L}_t$. We have:

$$\tilde{\phi}(\xi) := \frac{1}{2\pi i} \int\limits_{i\mathbb{R}} dz e^{\xi z} \phi(-iz\boldsymbol{\sigma}) \quad , \quad \phi(-iz) = \int\limits_{\mathbb{R}_+} d\xi e^{-\xi z} \tilde{\phi}(\xi) \qquad (54)$$

In the pure scaling case for $D = 1$ the inverse Laplace transform $\tilde{\phi}$ reads:

$$\tilde{\phi}(\xi) = \sigma^{D\mu} \frac{1}{2\pi i} \int\limits_{i\mathbb{R}} dz e^{\xi z} |z|^{D\mu} = \frac{\sigma^{D\mu}}{2\cos(D\mu\pi/2)} \left(I_{+,\xi}^{-D\mu} + I_{-,\xi}^{-D\mu} \right) [\delta](\xi)$$

$$\times \frac{\sigma^{D\mu}}{\Gamma(-D\mu)2\cos(\frac{\pi}{2}D\mu)} \frac{1}{\xi^{D\mu+1}} \qquad (55)$$

where $I_{\pm,x}^{-\mu} = \mathcal{D}_{\pm,x}^{\mu}$ and the later operators are Marchaud whole axis fractional derivatives $\phi(-iz\boldsymbol{\sigma})$ may be in general unbounded as a function of z and thus the quantity $\tilde{\phi}(\boldsymbol{\xi})$ is in general not a function but a functional.

Proposition 2. *The coefficients $\mathfrak{E}_n$ in (52) read:*

$$\mathfrak{E}_n = \begin{cases} (-1) \int\limits_{\mathbb{R}_+} d\xi \tilde{\phi}(\xi) \left(e^{-\xi} - 1 + \xi \right) & \text{if } n = 1 \\ (-1)^n \int\limits_{\mathbb{R}_+} d\xi \tilde{\phi}(\xi) \frac{\xi^n}{n!} & \text{if } n > 1 \end{cases} \qquad (56)$$

From (56) and (55) we see that the coefficients are infinite if $D\mu < 2$.

We proceed as follows to solve the PDE (52). In the definition (56) of the coefficients $\mathfrak{E}_n$ we truncate the upper limit of integration at some threshold value then we solve the generalized Black & Scholes equation (52) analytically by Fourier transforming with respect to $\log(S)$ and at the end we take the limit of the truncation threshold to infinity. Note that this step is essential. Indeed, as seen from (54) and from (35) it is not clear if the inverse Laplace transform $\tilde{\phi}$ related to the truncated Levy distribution diminishes fast enough away from the origin and thus if the integral in (56) exists. We accomplish this task in section (3.2). Prior to doing that we describe how we will compute the number stocks as follows.

We define a utility function U of the portfolio as a functional of the price of the stock viz:

$$U := \int\limits_0^T V(\xi) d\xi = \int\limits_0^T \left(N_S S_\xi + C(S; \xi) \right) d\xi \qquad (57)$$

and require (57) to be minimal. We do not investigate here the mathematical subtleties concerned with the existence of the stochastic integral (57). The necessary condition is that the variation δU with respect to the price of the stock functional is zero. We have:

$$\delta U := \int\limits_0^T \left(N_S + \frac{\partial C(S; \xi)}{\partial S} \right) \delta S d\xi = 0 \qquad (58)$$

what yields that

$$N_S = -\frac{\partial C(S;\xi)}{\partial S} \tag{59}$$

as in the Gaussian case. We note that this choice of the number of stocks ensures the self-financing property of the portfolio. Indeed in the Cox-Ross-Rubinstein binary tree model in discrete time one considers a portfolio composed of a stock and a bond and one derives the number of stocks by requiring contingent claim replication, meaning an equality of the portfolio and the claim with probability one (see [28] for example). The later result is essentially the same as that in (59).

Comments. We have derived a PDE for the option price that ensures that the derivative of the expectation value of the portfolio with compounded interest is zero

$$\lim_{dt \to 0} \frac{E\left[V(t+dt) - e^{rdt}V(t) \,|S_t\right]}{dt} = 0 \tag{60}$$

without making any assumptions about the relationship between the drift of the stock price α and the riskless rate of interest r. We differ in that from standard models in financial mathematics [20, 21], models that assume at the outset that $\alpha = r$.

3.2 Final Result

We solve the generalized Black & Scholes equation analytically. Inserting (59) into the second equality in (52) we get:

$$\partial_t C = rC - r\frac{\partial C}{\partial x} + \sum_{n=1}^{\infty} \mathfrak{E}_n \frac{\partial^n C}{\partial x^n} \tag{61}$$

where $x = \log(S_t)$. The coefficients $\mathfrak{E}_n$ are defined in (56) with the upper limit of integration being truncated at some threshold value. Since the coefficients do not depend on x the PDE (61) is converted into a Ordinary Differential Equation (ODE) by taking a Fourier transform of the option price with respect to x. This gives:

$$\partial_t \tilde{C}(k;t) = (r + H(k))\, \tilde{C}(k;t) \tag{62}$$

where

$$C(x;t) := (2\pi)^{-1} \int_{\mathbb{R}} dx \tilde{C}(k;t) e^{-\imath kx} \tag{63}$$

and

$$H(k) := \left[r\imath k + \sum_{n=1}^{\infty} \mathfrak{E}_n(-\imath k)^n \right] = r\imath k + \mathcal{V}(k) \tag{64}$$

We insert (56) into (64) and obtain the following expression for the function $\mathcal{V}(k)$ that we call 'the Hamiltonian' after Hagen Kleinert [16]. We have:

$$\mathcal{V}(k) = \int_{\mathbb{R}_+} d\xi \tilde{\phi}(\xi) \left[\left(-1 + e^{-\xi} \right) (\imath k) + e^{\imath k \xi} - 1 \right]$$

$$= (-1)^2 \int_{\mathbb{R}_+} d\xi I^{(2)}_{-,\xi} \left[\tilde{\phi} \right] (\xi) \left[(\imath k e^{-\xi}) - k^2 e^{\imath k \xi} \right] \tag{65}$$

We see that the integrals in (65) exist. Therefore the limit of the truncation threshold in these integrals going to infinity can be performed at this stage. This is what we do now and assume hereafter the whole positive real axis in the integration in (65). From (54) we obtain the Hamiltonian

$$\mathcal{V}(k) = (\imath k) \phi(-\imath \boldsymbol{\sigma}) + \phi(-k\boldsymbol{\sigma}) \tag{66}$$

Now we come back to equation (62) which we solve subject to an initial condition at maturity as follows:

$$\tilde{C}(k;t) = \tilde{C}(k;T) \exp\left\{ - (r + H(k)) \tau \right\} \tag{67}$$

where $\tilde{C}(k;T)$ is the Fourier transform of the option payoff $C(x;T)$ at maturity T and $\tau := T - t$ is the time to maturity. This payoff reads:

$$C(x;T) = \begin{cases} \max(e^x - K, 0) \text{ for a call} \\ \max(K - e^x, 0) \text{ for a put} \end{cases} \tag{68}$$

where K is the strike price. The Fourier transform of the payoff is easily computed and it reads:

$$\tilde{C}(k;T) := \int_{\mathbb{R}} \xi C(\xi;T) e^{\imath k \xi} = \begin{cases} K^{\imath k+1} \left(-\frac{1}{\imath k+1} + \frac{1}{\imath k} + 2\pi\delta(k - \imath) \right) \text{ for a call} \\ K^{\imath k+1} \left(+\frac{1}{\imath k+1} - \frac{1}{\imath k} - 2\pi\delta(k - \imath) \right) \text{ for a put} \end{cases} \tag{69}$$

We insert (69) into (67) and invert the Fourier transform for a call. We have:

$$C(x;t) := \frac{1}{2\pi} \int_{\mathbb{R}} dk \tilde{C}(k;t) e^{-\imath k x} \tag{70}$$

$$= S_t \int_{-\infty}^{-\log(m)+r\tau} d\xi e^{-\xi} \nu_{\boldsymbol{\sigma} \cdot \mathbf{L}_\tau}(\xi + \phi(-\imath \boldsymbol{\sigma})\tau) - Ke^{-r\tau}$$

$$\times \int_{-\infty}^{-\log(m)+r\tau} d\xi \nu_{\boldsymbol{\sigma} \cdot \mathbf{L}_\tau}(\xi + \phi(-\imath \boldsymbol{\sigma})\tau) \tag{71}$$

$$= S_t N_1(d_1) - Ke^{-r\tau} N_2(d_1) \tag{72}$$

We recall that here $\nu_{\boldsymbol{\sigma} \cdot \mathbf{L}_\tau}(\xi)$ is the probability density function of the fluctuation term $\boldsymbol{\sigma} \cdot \mathbf{L}_\tau$.

In (72) we have changed the integration variables and simplified the result. Here we defined:

$$N_1(d) := \int\limits_{-\infty}^{d} d\xi e^{-\xi} e^{\tau\phi(-\imath\sigma)} \nu_{\sigma \cdot \mathbf{L}_\tau}(\xi) \, ,$$

$$N_2(d) := \int\limits_{-\infty}^{d} d\xi \nu_{\sigma \cdot \mathbf{L}_\tau}(\xi) \text{ and}$$

$$d_1 = -\log(m) + \tau \left(r + \phi(-\imath\sigma)\right) \tag{73}$$

In the limit $D\mu \to 2_-$ the density $\nu_{\sigma \cdot \mathbf{L}_\tau}(\xi)$ in (71) goes into a Gaussian with mean zero and variance $2\sigma\tau$ and (72) goes into the Gaussian Black & Scholes equation, see e.g. eqs. (1.6),(1.7) on page 8 in [11]. We end this section by stating the price of the portfolio. We have:

$$V(t) = N_S S_t + C(x; t) = -K e^{-r\tau} N_2(d_1) \tag{74}$$

Since the last factor on the right-hand side in (74) depends implicitly on S_t the unconditional expectation value of the portfolio does not increase exponentially as required. Therefore the solution (72) is only an approximation. However since the Gaussian Black & Scholes equation is a particular case of (72) it turns out that it is also only an approximation.

The factors in (73) are complex which is of course unrealistic. The reason for that is the following. In our approach we assumed that the time change dt is infinitesimally small rather asumming it to be finite at the outset and taking the limit $dt \to 0$ at the end of the calculation. We have checked that the later procedure leads to a real result which has the same form as in (72) except that the Lévy density $\nu_{\sigma \cdot \mathbf{L}_\tau}(\xi)$ goes into an inverse Fourier transform of $\exp\left(1/2\sigma^\mu \Gamma(\imath k + \mu)/\Gamma(\imath k)\right)$ evaluated at $\xi - r\tau\phi(-\imath\sigma)$ which essentially amounts to replacing the expression $-k^\mu + \imath k$ by $\Gamma(\imath k + \mu)/\Gamma(\imath k)$ in some intermediate calculations. The inverse Fourier transform in question is essentially equal to the $\log(S_\tau) - \alpha\tau$ process probability density function evaluated at the argument $\xi - r\tau\phi(-\imath\sigma)$. Therefore the price of the option is a discounted present value of the maturity payoff under a risk-neutral probability measure where the measure in question is related to the compensated log-price process $log(S_t) - \alpha t$. Thus we have proven that the risk-neutral option pricing method holds in the generic setting of operator stable processes.

Expressions (73) are difficult to deal with in numerical calculations. Indeed the "typical width" of the Fourier transform of the Lévy density is $\sigma^\mu \tau$. Since this quantity is small, meaning of the order of 10^{-2} for stock daily data and for times to maturity of the order of hundreds of days, the use of "primitive" methods like Romberg quadratures for evaluating the Fourier integrals requires a very high precision of calculation that is much bigger than the precision of the estimated parameters. Therefore we propose to use a more sophisticated method for the numerical integrations. This method is described in the Appendix.

4 Conclusions

We have applied the technique of characteristic functions to the problem of pricing an option on a stock that is driven by operator stable fluctuations. We have developed a technique to ensure that the expectation value of the portfolio grows exponentially with time. In doing this we have not, unlike other authors, made any assumptions about the analytic properties of the log-characteristic function of the stock price process. Instead we have expressed all results in terms of the characteristic function of the operator stable fluctuation $\mathbf{L}_1$.

Subsequent to successful numerical tests, we ought then to be able to price analytically not only European options but also exotic options with a finite number of different exercise times. This should also allow us to price American style options by allowing the number of exercise times to become infinite.

We may also compute the 99th percentile of the probability distribution of the deviation of the portfolio (Value at Risk) as a function of σ and of the log-characteristic function ϕ of the random vector $\mathbf{L}_1$. The Value at Risk will be expressed as an integral equation involving the conditional characteristic function of the portfolio deviation (46). The resulting integrals will be carried out by means of the Cauchy complex integration theorem.

The results of these calculations will be reported in a future publication.

Acknowledgement. This work resulted from research conducted within the SFI Basic Research Grant 04/BR/0251. We are grateful to Mark Meerschaert, Stefan Thurner, Christoli Bieli and Krzysztof Urbanowicz for useful discussions.

5 Appendix

We explain how the integrals from the Levy density in (72) are computed numerically in the case $D = 1$. Note that in this case equation (72) can be written as follows:

$$C(x;t) = S_t N_{\sigma L_\tau}^{(1)}(d; z) - K e^{-r\tau} N_{\sigma L_\tau}^{(0)}(d; z) \tag{75}$$

where $z := z_r + \imath z_i = \phi(-\imath\sigma)\tau$ is a complex number, $d := -\log(m) + r\tau$ is a real number and

$$N_{\sigma L_\tau}^{(s)}(d; z) := \int\limits_{-\infty}^{d} d\xi e^{-s\xi} \nu_{\sigma L_\tau}(\xi + z) \tag{76}$$

for $s = 0, 1$. We note that the factors (76) have a following integral representation that lends itself to numerical computations in a straightforward manner. We have:

$$
N_{\sigma L_\tau}^{(s)}(d; z) = \int\limits_{-\infty}^{d} d\xi e^{-s\xi} \cdot \frac{1}{2\pi} \int\limits_{\mathbb{R}} dk e^{-ik(\xi+z)} \cdot e^{-\tau\phi(\sigma k)}
$$

$$
\underset{A\to\infty}{=} \frac{1}{2\pi} \int\limits_{\mathbb{R}} dk e^{-ikz} \cdot \int\limits_{-A}^{d} d\xi e^{-(s+ik)\xi} e^{-\tau\phi(\sigma k)} \tag{77}
$$

$$
\underset{A\to\infty}{=} \frac{1}{2\pi} \int\limits_{\mathbb{R}} dk e^{-ikz} \cdot \left(\frac{-e^{-i\theta d} + e^{i\theta A}}{i\theta} \right) \cdot e^{-\tau\phi(\sigma k)}
$$

$$
\underset{A\to\infty}{=} \frac{1}{2\pi} \int\limits_{\mathbb{R}+is} dk e^{-ikz} \cdot \left(\frac{-e^{-i\theta d} + e^{i\theta A}}{i\theta} \right) \cdot e^{-\tau\phi(\sigma k)} \tag{78}
$$

$$
\underset{A\to\infty}{=} \frac{1}{2\pi} \int\limits_{\mathbb{R}} d\theta e^{-i(\theta+is)z} \cdot \left(\frac{-e^{-i\theta d} + e^{i\theta A}}{i\theta} \right) \cdot e^{-\tau\phi(\sigma(\theta+is))} \tag{79}
$$

$$
= \frac{e^{sz}}{2} \left[e^{-\tau\phi(is\sigma)} + \frac{1}{\pi} \int\limits_{0}^{\infty} d\theta e^{\theta z i} \right.
$$

$$
\left. \times \left[\frac{\sin(\theta(d + z_r))}{\theta} \mathcal{M}_1(\theta) + i \frac{\cos(\theta(d + z_r))}{\theta} \mathcal{M}_2(\theta) \right] \right] \tag{80}
$$

In the first equality in (77) we expressed the Lévy stable density through its Fourier transform and in the second equality in (77) we changed to order of integration. In the first equality in (78) we integrated over ξ and we defined $\theta = k - is$ and in the second equality in (78) we shifted the integration line by is in the complex plane. In doing this we used the Cauchy theorem applied to a rectangular contour composed of an interval $[-R, R]$, of that interval shifted by is and of sections perpendicular to the real axis that complete the contour. In the limit $R \to \infty$ the integrals over the later sections vanish. In (79) we factorize-d the integral and in (80) we performed the limit $A \to \infty$ by using the identity:

$$
\lim_{A\to\infty} \frac{e^{i\theta A}}{i\theta} = \pi\delta(\theta) \quad \text{for} \quad \theta \in \mathbb{R} \tag{81}
$$

where

$$
\mathcal{M}_1(\theta) := \sum_{p=\pm 1} e^{-\tau\phi(\sigma(p\theta+is))} \quad \text{and} \quad \mathcal{M}_2(\theta) := \sum_{p=\pm 1} p e^{-\tau\phi(\sigma(p\theta+is))} \tag{82}
$$

In the pure scaling case in one dimension, from (16), we have $\phi(k) = \phi_\pm k^\mu$ and thus:

$$\mathcal{M}_1(\theta) := \sum_{p=\pm 1} e^{-\phi_{\pm}\tau(\sigma l)^\mu \cos(\mu\phi_p)} \left(\cos(\alpha_p) - \imath \sin(\alpha_p)\right) \quad \text{and}$$

$$\mathcal{M}_2(\theta) := \sum_{p=\pm 1} p e^{-\phi_{\pm}\tau(\sigma l)^\mu \cos(\mu\phi_p)} \left(\cos(\alpha_p) - \imath \sin(\alpha_p)\right) \tag{83}$$

where

$$l := \sqrt{\theta^2 + s^2}, \cos(\phi_p) = \frac{p\theta}{l}, \sin(\phi_p) = \frac{s}{l}, \quad \text{and} \quad \alpha_p = \phi_{\pm}\tau(\sigma l)^\mu \sin(\mu\phi_p) \tag{84}$$

Since, as seen from (82), $\mathcal{M}_1(0) = 2e^{-\tau\phi(\imath\sigma s)}$ and $\mathcal{M}_2(0) = 0$ the integral in (80) is clearly finite the result can be used for numerical calculations. In the Gaussian case $\mu = 2$ we have $\alpha_p = 2\phi_{\pm}\tau\sigma^2 sp\theta$ and thus

$$\mathcal{M}_1(\theta) := 2\cos(2\phi_{\pm}s\sigma^2\theta\tau)e^{-\phi_{\pm}\tau\sigma^2(\theta^2-s^2)} \quad \text{and}$$

$$\mathcal{M}_2(\theta) := -2\imath\sin(2\phi_{\pm}s\sigma^2\theta\tau)e^{-\phi_{\pm}\tau\sigma^2(\theta^2-s^2)} \tag{85}$$

and $z = \tau\phi(-\imath\sigma) = z_r + \imath z_i = -\tau\phi_{\pm}\sigma^2$. Inserting (85) into (80) gives:

$$N_{\sigma L_\tau}^{(s)}(d; z) = \frac{1}{2} + \frac{1}{2\pi} \int_{\mathbb{R}} d\theta e^{-\tau\phi_{\pm}(\sigma\theta)^2} \cdot \frac{\sin(\theta e)}{\theta}$$

$$= \frac{1}{2} + \frac{1}{2\pi} \int_{\mathbb{R}} d\theta e^{-\tau\phi_{\pm}(\sigma\theta)^2} \cdot \left(\frac{1}{2}\int_{-e}^{e} d\eta e^{-\imath\eta\theta}\right) \tag{86}$$

$$= \frac{1}{2} + \frac{1}{2} \int_{-e}^{e} d\eta \nu_{\sigma L_\tau}(\eta) \int_{-\infty}^{e} d\eta \nu_{\sigma L_\tau}(\eta)$$

$$= \int_{-\infty}^{\frac{e}{\sqrt{2\phi_{\pm}\sigma^2\tau}}} d\eta \frac{1}{\sqrt{2\pi}} \exp\left\{-\frac{1}{2}\eta^2\right\} \tag{87}$$

where $e = -\log(m) + \tau\left(r + \phi_{\pm}\sigma^2(2s-1)\right)$. From (87) we see that the factors coincide with those in the Gaussian Black & Scholes formula.

References

1. Bachelier L (1900), Theory of Speculation, Ann. Sci. Ecole Norm. Sup. 3: 21; preprint from P.H. Cootner (editor), The random character of stock prices, second edition (MIT Press Cambridge, 1969).
2. Gopikrishnan P et al (1998) Inverse cubic law for the distribution of stock price variations, Eur. Phys. J. B 3: 139–140.
3. Mandelbrot B (1963) The variations of certain speculative prices, J. of Business 36: 392–417

4. Fama EF (1970) Efficient Capital Markets: A Review of Theory and Empirical Work, J. of Finance 25: 383–417

5. Meerschaert MM, Scheffler HP (2003) Portfolio modeling with heavy tailed random vectors, in Handbook of Heavy-Tailed Distributions in Finance, Rachev ST, Ed., 595–640, Elsevier North-Holland, New York.

6. Rachev S, Mittnik S (2000) Stable Paretian Models in Finance, Wiley, Chichester.

7. Meerschaert MM, Scheffler HP (2001) Limit Distributions for Sums of Independent Random Vectors: Heavy tails in Theory and Practice. John Wiley & Sons, Inc.

8. Emmer S, Kleuppenberg C (2004) Optimal portfolios when stock prices follow an exponential Lévy process Finance and Stochastics 8: 17–44

9. Nikias CL, Shao M (1995) Signal Processing with Alpha-Stable Distributions and Applications, New York, John Wiley and Sons.

10. Dash Jan W, Path Integrals and Options - I, preprint available on-line at http://www.physik.fu-berlin.de/~kleinert/b3/papers/ by courtesy, Kleinert H.

11. Rama C, Integro-differential equations and numerical methods, in:*Financial Modeling with Jump Processes* Chapman & Hall, CRC Financial Mathematics Series, 381–430

12. Bertoin J (1996) Lévy processes as Markov processes, in: Lévy processes Cambridge University Press.

13. Boyarchenko SI, Levendorskii SZ (2005) General Option Exercise Rules, with Applications to Embedded Options and Monopolistic Expansion (October 30) Available at SSRN: http://ssrn.com/abstract=838624

14. Samko SG, Kilbas AA, Marichev OI (1993) Fractional Integrals and Derivatives Theory and Applications, Gordon and Breach Science Publishers S.A.

15. Dzherbashyan MM, Nersesyan AB, The criterion of the expansion of the functions to the Dirichlet series, Izv. Akad. Nauk Armyan. SSR Ser. Fiz.-Mat. Nauk, 11: no 5. 85–108

16. Kleinert H (2004) Option Pricing for Gaussian, for non-Gaussian fluctuations, and for a fluctuating variance, chapters 20.4.3 – 20.4.5, 1416–1428 in: Path Integrals in Quantum Mechanics, Statistics, Polymer Physics, and Financial Markets, World Scientific Publishing Co., Singapore 3 ed.

17. Redner S (2001) A guide to first passage processes, Cambridge University Press.

18. Kleinert H (2004) Path Integrals in Quantum Mechanics, Statistics, Polymer Physics, and Financial Markets, World Scientific Publishing Co., Singapore, 3 ed.

19. Kleinert H (2002) Option Pricing from Path Integral for Non-Gaussian Fluctuations. Natural Martingale and Applications to Truncated Lévy Distributions, preprint cond-mat/0202311

20. Cont R, Tankov P (2004) Risk neutral modelling with exponential Lévy processes, 353–379 in: Financial Modelling with Jump Processes, Chapman & Hall, Financial Mathematics Series.

21. Hurst SR, Platen E, Rachev ST (1999) Option Pricing for a LogStable Asset Pricing Model, Mathematical and Computer Modelling 29: 105-119.

22. Rachev S, Mittnik S (2000) Stable Paretian Models in Finance, John Wiley & Sons.

23. Cartea Á, Howinson S, Distinguished Limits of Lévy Stable Processes, and Applications to Option Pricing, Oxford Financial Research Centre, No 2002mf04.

24. McCulloch HJ (2004), The Risk-Neutral Measure and Option Pricing under Log-Stable Uncertainty, Econometric Society 2004 North American Winter Meetings 428, Econometric Society.

25. Zolotarev VM (1957) Mellin-Stieltjes Transforms in Probability Theory, Theory Prob. Appl., 2: No. 4, 433-460.

26. Cambanis S, Miller G (1981) Linear Problems in pth order and Stable Processes, SIAM J. Appl. Math., 41: (Aug.), 43–69.

27. Wolfe SJ (1973) On the Local Behavior of Characteristic Functions, Ann. Prob., 1: No. 5, 862–866.

28. Musiela M, Rutkowski M (1998) The Cox-Ross-Rubinstein Model, in: Martingale Methods in Financial Modelling, Springer.

29. Ibragimov IA, Chernin KE (1959), On the unimodality of Stable laws, Theory of Probability and its Applications, 4: No 4, 417–19

Inferring the Composition
of a Trader Population in a Financial Market

Nachi Gupta[1], Raphael Hauser[1], and Neil F. Johnson[2]

[1] Oxford University Computing Laboratory, Numerical Analysis Group,
Wolfson Building, Parks Road, Oxford OX1 3QD, U.K.
nachi@comlab.ox.ac.uk
[2] Oxford University, Department of Physics, Clarendon Building, Parks Road,
Oxford OX1 3PU, U.K.

1 Introduction

There has been an explosion in the number of models proposed for understanding and interpreting the dynamics of financial markets. Broadly speaking, all such models can be classified into two categories: (a) models which characterize the macroscopic dynamics of financial prices using time-series methods, and (b) models which mimic the microscopic behavior of the trader population in order to capture the general macroscopic behavior of prices. Recently, many econophysicists have trended towards the latter by using multi-agent models of trader populations. One particularly popular example is the so-called Minority Game [1], a conceptually simple multi-player game which can show non-trivial behavior reminiscent of real markets. Subsequent work has shown that – at least in principle – it is possible to train such multi-agent games on real market data in order to make useful predictions [2–5]. However, anyone attempting to model a financial market using such multi-agent trader games, with the objective of then using the model to make predictions of real financial time-series, faces two problems: (a) How to choose an appropriate multi-agent model? (b) How to infer the level of heterogeneity within the associated multi-agent population?

This paper addresses the question of how to infer the multi-trader heterogeneity in a market (i.e. question (b)) assuming that the Minority Game, or one of its many generalizations [2, 3], forms the underlying multi-trader model. We consider the specific case where each agent possesses a pair of strategies and chooses the best-performing one at each timestep. Our focus is on the uncertainty for our parameter estimates. Using real financial data for quantifying this uncertainty, represents a crucial step in developing associated risk measures, and for being able to identify pockets of predictability in the price-series.

As such, this paper represents an extension of our preliminary study in [6]. In particular, the present analysis represents an important advance in that it generalizes the use of probabilities for describing the agents' heterogeneity. Rather than using a probability, we now use a finite measure, which is not necessarily normalized to unit total weight. This generalization yields a number of benefits such as a stronger preservation of positive definiteness in the covariance matrix for the estimates. In addition, the use of such a measure removes the necessity to scale the time-series, thereby reducing possible further errors. We also look into the problem of estimating the finite measure over a space of agents which is so large that the estimation technique becomes computationally infeasible. We propose a mechanism for making this problem more tractable, by employing many runs with small subsets chosen from the full space of agents. The final tool we present here is a method for removing bias from the estimates. As a result of choosing subsets of the full agent space, an individual run can exhibit a bias in its predictions. In order to estimate and remove this bias, we propose a technique that has been widely used with Kalman Filtering in other application domains.

2 The Multi-Agent Market Model

Many multi-agent models – such as the Minority Game [1] and its generalizations [2,3] – are based on binary decisions. Agents compete with each other for a limited resource (e.g. a good price) by taking a binary action at each time-step, in response to global price information which is publicly available. At the end of each time-step, one of the actions is denoted as the winning action. This winning action then becomes part of the information set for the future. As an illustration of the tracking scheme, we will use the Minority Game – however we encourage the reader to choose their own preferred multi-agent game. The game need not be a binary-decision game, but for the purposes of demonstration we will assume that it is.

2.1 Parameterizing the Game

We provide one possible way of parameterizing the game in order to fit the proposed methodology. We select a time horizon window of length T over which we score strategies for each agent. This is a sliding window given by $(w_{k-T}, \ldots, w_{k-1})$ at time step k. Here $w_k = -\text{sgn}(z_k)$ represents what would have been the winning decision at time k, and z_k is the difference in the corresponding price-series, or exchange-rate, r_k.

$$z_k = r_k - r_{k-1} \tag{1}$$

Each agent has a set of strategies which it scores on this sliding time-horizon window at each time-step. The agent chooses its highest scoring strategy as its winning strategy, and then plays it. Assume we have N such agents. At each time-step they each play their winning strategy, resulting in a global outcome for the game at that time-step. Their aggregate actions result in an outcome which we expect to be indicative of the next price-movement in the real price-series. If one knew the population of strategies in use, then one could predict the future price-series with certainty – apart from a few occasions where ties in strategies might be broken randomly.

The next step is to estimate the heterogeneity of the agent population itself. We choose to use a recursive optimization scheme similar to a Kalman Filter – however we will also force inequality constraints on the estimates so that we cannot have a negative number of agents of a given type playing the game. Suppose x_k is the vector at time-step k representing the heterogeneity among the N types of agents in terms of their strategies. We can write this as

$$x_k = \begin{bmatrix} x_{1,k} \\ \vdots \\ x_{N,k} \end{bmatrix} \tag{2}$$

where we force each element of the vector to be ≥ 0.

$$x_{i,k} \geq 0, \forall i \tag{3}$$

We provided a very similar scheme recently in [6], in which we had further constrained our estimate to a probability space and as a result also had to re-scale the time-series to better span the interval $[-1, 1]$. We now relax the constraint condition (and the re-scaling) since the benefits of lying within the probability space are outweighed by the benefits of allowing the estimate to move out of this space. One significant benefit of staying within the probability space is the ability to bound covariance matrices on errors (since upper bounds on the probability of certain events are known). On the other hand, staying constrained to a probability space removes one degree of freedom from our system (i.e. $x_{N,k} = 1 - x_{1,k} - \cdots - x_{n-1,k}$). This can cause the covariance of our estimate to become ill-formed by a possible numerical loss of positive definiteness (which we could prevent by artificially inflating the diagonal elements of the covariance matrix).

3 Recursive Optimization Scheme

In the following subsections we introduce the Kalman Filter, after which we will discuss some desirable extensions for our work.

3.1 Kalman Filter

A Kalman Filter is a recursive least-squares implementation, which makes only one pass through the data such that it can wait for each measurement to come in real time, and then make an estimate for that time given all the information from the past. The Kalman Filter holds a minimal amount of information in its memory at each time, yielding a relatively cheap computational cost for solving the optimization problem. In addition, the Kalman Filter can make a forecast n steps ahead and provides a covariance structure concerning this forecast. The Kalman Filter is a predictor-corrector system, that is to say, it makes a prediction and upon observation of real data, it perturbs the prediction slightly as a correction, and so forth.

The Kalman Filter attempts to find the best estimate at every iteration, for a system governed by the following model:

$$x_k = F_{k,k-1} x_{k-1} + u_{k,k-1}, \qquad u_{k,k-1} \sim N(0, Q_{k,k-1}) \tag{4}$$

$$z_k = H_k x_k + v_k, \qquad v_k \sim N(0, R_k) \tag{5}$$

Here x_k represents the true state of the underlying system, which in our case is the finite measure over the agents. $F_{k,k-1}$ is a matrix used to make the transition from state x_{k-1} to x_k. In our applications, we choose $F_{k,k-1}$ to be the identity matrix for all k since we assume that locally the finite measure over the strategies doesn't change drastically. It would be a tough modeling problem to choose another matrix (i.e., not the identity matrix) – however, if desired we could incorporate a more complex transition matrix into the model, even one that is dependent on previous outcomes. The variable z_k represents the measurement (also called observation). H_k is a matrix that relates the state space and measurement space by transforming a vector in the state space to the appropriate vector in the measurement space. For our artificial market model, H_k will be a row vector containing the decisions based on each of the agent's winning strategies. So x_k, the measure over the agents, acts as a weighting on the decisions for each agent, and the inner product $H_k x_k$ can be thought of as a weighted average of the agents' decisions – this represents the aggregate decision made by the system of agents. The variables $u_{k,k-1}$ and v_k are both noise terms which are normally distributed with mean 0 and variances $Q_{k,k-1}$ and R_k, respectively.

The Kalman Filter will at every iteration make a prediction for x_k, which we denote by $\hat{x}_{k|k-1}$. We use the notation $k|k-1$ since we will only use measurements provided until time-step $k-1$ in order to make the prediction at time k. We can define the state prediction error $\tilde{x}_{k|k-1}$ as the difference between the true state and the state prediction.

$$\tilde{x}_{k|k-1} = x_k - \hat{x}_{k|k-1} \tag{6}$$

In addition, the Kalman Filter will provide a state estimate for x_k, given all the measurements provided up to and including time step k. We denote these estimates by $\hat{x}_{k|k}$. We can similarly define the state estimate error by

$$\tilde{x}_{k|k} = x_k - \hat{x}_{k|k} \tag{7}$$

Since we assume $u_{k,k-1}$ is normally distributed with mean 0, we make the state prediction simply by using $F_{k,k-1}$ to make the transition. This is given by

$$\hat{x}_{k|k-1} = F_{k,k-1}\hat{x}_{k-1|k-1} \qquad (8)$$

We can also calculate the associated covariance for the state prediction, which we call the covariance prediction. This is actually just the expectation of the outer product of the state prediction error with itself. This is given by

$$P_{k|k-1} = F_{k,k-1}P_{k-1|k-1}F'_{k,k-1} + Q_{k,k-1} \qquad (9)$$

Notice that we use the prime notation on a matrix throughout this paper to denote the transpose. Now we can make a prediction on what we expect to see for our measurement, which we call the measurement prediction, by

$$\hat{z}_{k|k-1} = H_k\hat{x}_{k|k-1} \qquad (10)$$

The difference between our true measurement and our measurement prediction is called the measurement residual, which we calculate by

$$\nu_k = z_k - \hat{z}_{k|k-1} \qquad (11)$$

We can also calculate the associated covariance for the measurement residual, which we call the measurement residual covariance, by

$$S_k = H_k P_{k|k-1} H'_k + R_k \qquad (12)$$

We now calculate the Kalman Gain, which lies at the heart of the Kalman Filter. This essentially tells us how much we prefer our new observed measurement over our state prediction. We calculate this by

$$K_k = P_{k|k-1}H'_k S_k^{-1} \qquad (13)$$

Using the Kalman Gain and measurement residual, we update the state estimate. If we look carefully at the following equation, we are essentially taking a weighted sum of our state prediction with the Kalman Gain multiplied by the measurement residual. So the Kalman Gain is telling us how much to 'weight in' information contained in the new measurement. We calculate the updated state estimate by

$$\hat{x}_{k|k} = \hat{x}_{k|k-1} + K_k\nu_k \qquad (14)$$

Finally, we calculate the updated covariance estimate. This is just the expectation of the outer product of the state error estimate with itself. Here we will give the most numerically stable form of this equation, as this form prevents loss of symmetry and best preserves positive definiteness

$$P_{k|k} = (I - K_kH_k)P_{k|k-1}(I - K_kH_k)' + K_kR_kK_k^T \qquad (15)$$

The covariance matrices throughout the Kalman Filter give us a way to measure the uncertainty of our state prediction, state estimate, and the measurement residual. Also notice that the Kalman Filter is recursive, and we require an initial estimate $\hat{x}_{0|0}$ and associated covariance matrix $P_{0|0}$. Here we simply provide the equations of the Kalman Filter without derivation. For a detailed description of the Kalman Filter, see Ref. [7].

3.2 Nonlinear Equality Constraints

As we are estimating a vector in which each element has a non-negative value, we would like to force the Kalman Filter to have some inequality constraints. We now introduce a generalization for nonlinear equality constraints followed by an extension to inequality constraints. In particular, let's add to our model (Eqs. (4) and (5)) the following smooth nonlinear equality constraints

$$e_k(x_k) = 0 \tag{16}$$

The constraints provided in Eq. (3) are actually linear. We present the nonlinear case for further completeness here. We now rephrase the problem we would like to solve, using the superscript c to denote constrained. We are given the last prediction and its covariance, the current measurement and its covariance, and a set of equality constraints and would like to make the current prediction and find its covariance matrix. Let's write the problem we are solving as

$$z_k^c = h_k^c(x_k) + v_k^c, \qquad v_k^c \sim N(0, R_k^c) \tag{17}$$

Here z_k^c, h_k^c, and v_k^c are all vectors, each having three distinct parts. The first part will represent the prediction for the current time step, the second part is the measurement, and the third part is the equality constraint. z_k^c effectively still represents the measurement, with the prediction treated as a "pseudo-measurement" with its associated covariance.

$$z_k^c = \begin{bmatrix} F_{k,k-1}\hat{x}_{k-1|k-1} \\ z_k \\ 0 \end{bmatrix} \tag{18}$$

The matrix h_k^c takes our state into the measurement space as before

$$h_k^c(x_k) = \begin{bmatrix} x_k \\ H_k x_k \\ e_k(x_k) \end{bmatrix} \tag{19}$$

Notice that by combining Eqs. (6) and (7), we can rewrite the state error prediction as

$$\tilde{x}_{k|k-1} = F_{k,k-1}\tilde{x}_{k-1|k-1} + u_{k,k-1} \tag{20}$$

We can define v_k^c again as the noise term using Eq. (20).

$$v_k^c = \begin{bmatrix} -F_{k,k-1}\tilde{x}_{k-1|k-1} - u_{k,k-1} \\ v_k \\ 0 \end{bmatrix} \tag{21}$$

v_k^c will be normally distributed with mean 0 and variance R_k^c. The diagonal elements of R_k^c represent the variance of each element of v_k^c. We define the covariance of the state estimate error at time-step k as $P_{k|k}$. Notice also that R_k^c contains no off-diagonal elements.

$$R_k^c = \begin{bmatrix} F_{k,k-1}P_{k-1|k-1}F_{k,k-1}' + Q_{k,k-1} & 0 & 0 \\ 0 & R_k & 0 \\ 0 & 0 & 0 \end{bmatrix} \tag{22}$$

This method of expressing our problem can be thought of as a fusion of (a) the state prediction, and (b) the new measurement at each iteration, under the given equality constraints. As we did for the Kalman Filter, we will state the equations here. The interested reader is referred to Refs. [8, 9].

$$\hat{x}_{k|k,j} = \begin{bmatrix} 0 & I \end{bmatrix} \begin{bmatrix} R_k^c & H_{k,j}^c \\ H_{k,j}^c{}' & 0 \end{bmatrix}^+ \begin{bmatrix} z_k^c - h_k^c(\hat{x}_{k|k,j-1}) + H_{k,j}^c \hat{x}_{k|k,j-1} \\ 0 \end{bmatrix} \tag{23}$$

Throughout this paper, we use the $+$ notation on a matrix to denote the pseudo-inverse. In this method we are iterating over a dummy variable j within each time-step, until we fall within a predetermined convergence bound $|\hat{x}_{k|k,j} - \hat{x}_{k|k,j-1}| \leq c_k$ or hit a chosen number of maximum iterations. We initialize our first iteration as $\hat{x}_{k|k,0} = \hat{x}_{k-1|k-1}$ and use the final iteration as $\hat{x}_{k|k} = \hat{x}_{k|k,J}$ where J represents the final iteration. Notice that we allowed the equality constraints to be nonlinear. As a result, we define $H_{k,j}^c = \frac{\partial h_k^c}{\partial x_k}(\hat{x}_{k|k,j-1})$ which gives us a local approximation to the direction of h_k^c.

A stronger form of these equations can be found in Refs. [8, 9], where R_k^c will reflect the tightening of the covariance for the state prediction based on the new estimate at each iteration of j. We do not use this form and tighten the covariance matrix within these iterations since in the next section we will require the flexibility of changing the number of equality constraints between iterations of j. By not tightening the covariance matrix in this way, we are left with a larger covariance matrix for the estimate (which shouldn't harm us significantly). This covariance matrix is calculated as

$$P_{k|k,j} = -\begin{bmatrix} 0 & I \end{bmatrix} \begin{bmatrix} R_k^c & H_{k,j}^c \\ H_{k,j}^c{}' & 0 \end{bmatrix}^+ \begin{bmatrix} 0 \\ I \end{bmatrix} \tag{24}$$

Notice that for faster computation times, we need only calculate $P_{k|k,j}$ for the final iteration of j. Further, if our equality constraints are in fact independent of j, we only need to calculate $H^c_{k,j}$ once for each k. This also implies that the pseudo-inverse in Eq. (23) can be calculated only once for each k.

This method, while very different from the Kalman Filter presented earlier, provides us with an estimate $\hat{x}_{k|k}$ and a covariance matrix for the estimate $P_{k|k}$ at each time-step, in a similar way to the Kalman Filter. However, this method allows us to incorporate equality constraints.

3.3 Nonlinear Inequality Constraints

We will now extend the equality constrained problem to an inequality constrained problem. To our system given by equations (4), (5), and (16), we will add the smooth inequality constraints given by

$$l_k(x_k) \geq 0. \tag{25}$$

Our method will be to keep a subset of the inequality constraints active at any time. An active constraint is simply a constraint that we treat as an equality constraint. We will ignore any inactive constraint when solving our optimization problem. After solving the problem, we then check if our solution lies in the space given by the inequality constraints. If it doesn't, we start from the solution in our previous iteration and move in the direction of the new solution until we hit a set of constraints. For the next iteration, this set of constraints will be the new active constraints.

We formulate the problem in the same way as before, keeping Eqs. (17), (18), (21), and (22) the same to set up the problem. However, we replace Eq. (19) by

$$h^c_k(x_k) = \begin{bmatrix} x_k \\ H_k x_k \\ e_k(x_k) \\ l^a_{k,j}(x_k) \end{bmatrix} \tag{26}$$

$l^a_{k,j}$ represents the set of active inequality constraints. Although we keep Eqs. (18), (21), and (22) the same, these will need to be padded by additional zeros appropriately to match the size of $l^a_{k,j}$. Now we solve the equality constrained problem consisting of the equality constraints and the active inequality constraints (which we treat as equality constraints) using Eqs. (23) and (24). Let's now call the solution from Eq. (23) $\hat{x}^*_{k|k,j}$ since we have not yet checked if this solution lies in the inequality constrained space. In order to check this, we find the vector that we moved along to reach $\hat{x}^*_{k|k,j}$. This is simply

$$d = \hat{x}^*_{k|k,j} - \hat{x}_{k|k,j-1} \tag{27}$$

We now iterate through each of our inequality constraints, to check if they are satisfied. If they are all satisfied, we choose $t_{\max} = 1$. If they are not, we choose the largest value of $t_{\max}$ such that $\hat{x}_{k|k,j-1} + t_{\max}d$ lies in the inequality constrained space. We choose our estimate to be

$$\hat{x}_{k|k,j} = \hat{x}_{k|k,j-1} + t_{\max}d \tag{28}$$

We also would like to remember the inequality constraints which are being touched in this new solution. These constraints will now become active for the

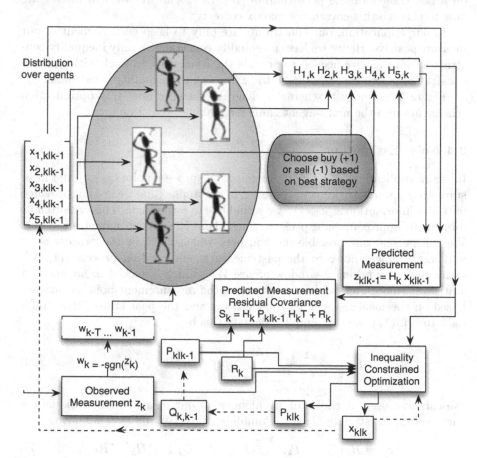

Fig. 1. Summary of the recursive method for predicting the heterogeneity of the multi-agent population. We have dropped the ^ notation. Shown is a situation with 5 types of agents, where each type has more than one strategy. They each score their strategies over the sliding time-horizon window $(w_{k-T} \cdots w_{k-i})$ and choose the best one. H_k represents the decisions they each make in this time-step, which in the case of a binary-decision game is $+1$ or -1. Taking the dot product of the frequencies over the agents and their decisions, we arrive at our prediction for the measurement. We then allow the recursion into the optimization technique. Since we chose $F_{k,k-1}$ as the identity matrix for all k, we omitted it entirely from this diagram. We also assume initial conditions are provided. In the next subsection, we describe how we arrive at the noise parameters $Q_{k,k-1}$ and R_k, which appear in the diagram.

next iteration and lie in $l^a_{k,j+1}$. Note that $l^a_{k,0} = l^a_{k-1,J}$, where J represents the final iteration of a given time-step. We do not perturb the error covariance matrix from Eq. (24) in any way. Under the assumption that our model is a well-matched model for the data, enforcing inequality constraints (as dictated by the model) should only make our estimate better. Having a slightly larger covariance matrix is better than having an overly optimistic one based on a bad choice for the perturbation [10]. In the future, we will investigate how to perturb this covariance matrix correctly.

In our application, our constraints are only to keep each element of our measure positive. Hence we have no equality constraints – only inequality constraints. However, we needed to provide the framework to work with equality constraints before we could make the extension to inequality constraints.

Figure 1 provides a schematic diagram showing how this optimization scheme fits into the multi-agent game for making predictions.

3.4 Noise Estimation

In many applications of Kalman Filtering, the process noise $Q_{k,k-1}$ and measurement noise R_k are known. However, in our application we are not provided with this information a priori so we would like to estimate it. This can often be difficult to approximate, especially when there is a known model mismatch. We will present one possible method here which matches the process noise and measurement noise to the past measurement residual process [11]. We estimate R_k by taking a window of size W_k (which is picked in advance for statistical smoothing) and time-averaging the measurement noise covariance based on the measurement residual process and the past states. If we refer back to Eq. (12), we can simply calculate this by

$$\hat{R}_k = \frac{1}{W_k - 1} \sum_{j=k-W_k}^{k-1} \nu_j \nu_j{}' - H_j P_{j|j-1} H_j' \qquad (29)$$

We can now use our choice of R_k along with our measurement residual covariance S_k, to estimate $Q_{k,k-1}$. Combining Eqs. (9) and (12) we have

$$S_k = H_k(F_{k,k-1} P_{k-1|k-1} F_{k,k-1}{}' + Q_{k,k-1}) H_k' + R_k \qquad (30)$$

Bringing all $Q_{k,k-1}$ terms to one side leaves us with

$$H_k Q_{k,k-1} H_k' = S_k - H_k F_k P_{k-1|k-1} F_k' H_k' - R_k \qquad (31)$$

Solving for $Q_{k,k-1}$ gives us

$$\hat{Q}_{k,k-1} = \left(H_k' H_k\right)^+ H_k' \left(S_k - H_k F_k P_{k-1|k-1} F_k' H_k' - R_k\right) H_k \left(H_k' H_k\right)^+ \qquad (32)$$

Note that it may be desirable to keep $\hat{Q}_{k,k-1}$ diagonal if we do not believe the process noise has any cross-correlation. It is rare that one would expect a cross-correlation in the process noise. In addition, keeping the process noise diagonal has the effect of making our covariance matrix 'more positive definite'. This can be done simply by setting the off-diagonal terms of $\hat{Q}_{k,k-1}$ equal to 0. It is also important to keep in mind that we are estimating covariance matrices here which must be symmetric and positive semidefinite, and the diagonal elements should always be greater than or equal to zero since these are variances.

4 Estimation in the Presence of an Ecology of Many Agent Types

It is very likely that we will come across multi-agent markets, and hence models, with many different agent types, e.g., $N > 100$. As N grows, not only does our state space grow linearly, but our covariance space will grow quadratically. We quickly reach areas where we may no longer be in a computationally feasible region. For example, if we look at strategies for agents playing the Minority Game and define a type of agent to have exactly 2 strategies, we see that as we increase the memory sizes of our agents our full set of pairs of strategies grows very quickly in relation to the memory size (e.g., $m = 1$ yields $2^{2^1} = 4$ strategies and $\binom{4}{2} = 6$ pairs of strategies, $m = 2$ yields $2^{2^2} = 16$ strategies and $\binom{16}{2} = 120$ pairs of strategies, $m = 3$ yields $2^{2^3} = 256$ strategies and $\binom{256}{2} = 32640$ pairs of strategies, $m = 4$ yields $2^{2^4} = 65536$ strategies and $\binom{65536}{2} = 2147450880$ pairs of strategies, ...). If we were interested in simultaneously allowing all possible pairs of strategies, our vectors and matrices for these computations would have a dimension that would not be of reasonable complexity, especially in situations where real-time computations are needed. In such situations, we propose selecting a subset of the full set of strategies uniformly at random, and choosing these as the only set that could be in play for the time-series. We can then do this a number of times and average over the predictions and their covariances. We would hope that this would cause a smoothing of the predictions and remove outlier points. In addition we might notice certain periods that are generally more predictable by doing this, which we call *pockets of predictability*.

4.1 Averaging over Multiple Runs

For each run j of our M runs, we have our predicted measurement at time k given by $\hat{z}_{k,j}$ and our predicted covariance for the measurement residual as $S_{k,j}$. Using the predicted measurements, we can simply average to find our best estimate of the prediction.

$$\hat{z}_k^* = \sum_{j=1}^{M} \frac{\hat{z}_{k,j}}{M} \tag{33}$$

Similarly, we can calculate our best estimate of the predicted covariance for the measurement residual:

$$S_k^* = \sum_{j=1}^{M} \frac{S_{k,j}}{M} \tag{34}$$

It is important to note here that since $\hat{z}_k^*$ and S_k^* are both estimators, as M tends to ∞, we expect the standard error of the mean for both to tend towards 0. Also, note that we chose equal weights when calculating the averages; we could have alternatively chosen to use non-equal weights had we developed a system for deciding on the weights.

4.2 Bias Estimation

Since we are choosing subsets of the full strategy space, we expect that in some runs a number of the strategies might tend to behave in the same way. This doesn't mean that the run is useless and provides no information. In fact, it could be the case that the run provides much information – it is just that the predictions always tend to be biased in one direction or the other. So what we might like to do is remove bias from the system. The simplest way to do this is to augment the Kalman Filter's state space with a vector of elements representing possible bias [12]. We can model this bias as lying in the state space, the measurement space, or some combination of elements of either or both. We redefine the model for our problem as

$$x_k^b = F_{k,k-1}^b x_{k-1}^b + u_{k,k-1}, \qquad u_{k,k-1} \sim N\left(0, Q_{k,k-1}^b\right) \tag{35}$$

$$z_k = H_k^b x_k^b + v_k, \qquad v_k \sim N\left(0, R_k\right) \tag{36}$$

where x_k^b represents the augmented state and b_k is the bias vector at time step k

$$x_k^b = \begin{bmatrix} x_k \\ b_k \end{bmatrix} \tag{37}$$

The transition matrix must also be augmented to match the augmented state. In the top left corner, we place our original transition matrix, and in the top right corner we place $B_{k,k-1}$ representing how the estimated bias term should be added into the dynamics. In the bottom left we have the zero matrix so the bias term is not dependent on the state x_k, and in the bottom right we have the identity matrix indicating that the bias is updated by itself exactly at each time.

$$F_{k,k-1}^b = \begin{bmatrix} F_{k,k-1} & B_{k,k-1} \\ 0 & I \end{bmatrix} \tag{38}$$

Similarly, we horizontally augment our measurement matrix, where C_k represents how the bias terms should be added into the measurement space.

$$H_k^b = \begin{bmatrix} H_k & C_k \end{bmatrix} \tag{39}$$

For the process noise, we keep the off diagonal elements as 0, assuming no cross-correlations between the state and the bias. We also generally assume no noise in the bias term and keep its noise covariance as 0, as well. Of course, this can be changed easily enough if the reader would like to model the bias with some noise.

$$Q^b_{k,k-1} = \begin{bmatrix} Q_{k,k-1} & 0 \\ 0 & 0 \end{bmatrix} \tag{40}$$

We can take this bias model framework and place it into the inequality constrained filtering scheme provided earlier, with the model given by Eq. (17), where we simply use the augmented states when necessary, rather than the regular state space (e.g. let $x_k = x^b_k$).

5 Example for a Foreign-Exchange Rate Series

We now apply these ideas to a data set of hourly USD/YEN foreign exchange rate data, from 1993 to 1994, using the Minority Game. We do not claim that this is a good model for the time-series, but it does contain some of the characteristics we might expect to see in this time-series. We look at all pairs of strategies with memory size $m = 4$ of which there are 2,147,450,880 as our set of possible types for agents. Since the size of this computation would not be tractable, we take a random subset of 5 of these types and use these 5 as the only possible types of agents to play the game. We perform 100 such runs, each time choosing 5 types at random. In addition, we allow for a single bias removal term. We could have many more terms for the bias, but we only use 1 in order to limit the growth of the state space. We assume the entire bias lies in a shifting of the measurements, so we don't use a $B_{k,k-1}$ from Eq. (38) and only choose C_k in Eq. (39) to be the identity matrix – or in our case simply 1 since C_k is 1x1.

For the analysis of how well our forecasts perform, we calculate the residual log returns and plot these. Given our time-series, we can calculate the log return of the exchange rate as $l_k = \log(r_k) - \log(r_{k-1})$. Note that based on our definition for z_k from Eq. (1), we can write the log return also as $l_k = \log(r_{k-1} + z_k) - \log(r_{k-1})$. Similarly, we can define our predicted forecast for the log return as $\hat{l}_k = \log(r_{k-1} + \hat{z}_k) - \log(r_{k-1})$. Given these two quantities, we can calculate the residual of the predicted log return and the observed log return as $\tilde{l}_k = l_k - \hat{l}_k$. Using the delta method [13], we can also calculate the variance of this residual to be $\frac{S_k}{(r_{k-1})^2}$. We perform 100 such runs, which we average over using the method described in Sect. 4.1. A good test for whether our variances are overly optimistic is to check if the measure satisfies the Chebyshev Inequality. For example, we can check visually that no more than about $\frac{1}{9}$ of the residuals lie within 3σ's. We plot the residual log return along with 3σ bounds and the standard error of the mean in Fig. 2. Visually, we would say the residuals in Fig. 2 would certainly satisfy this condition if they were centered about 0. Maybe further bias removal would readily achieve this.

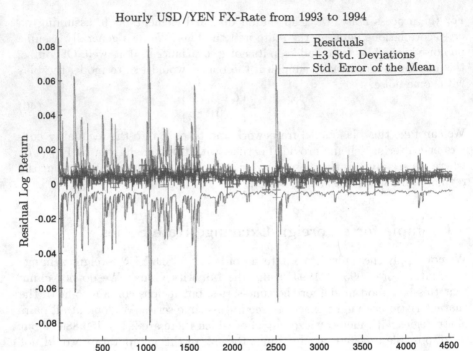

Fig. 2. We show the residuals of the log returns plotted with 3σ's centered about 0 and the standard error of the mean over the 100 runs. Despite the one parameter bias removal, we still see a general bias in the data without which the residuals look much cleaner. Perhaps a more complex bias model would remove this. Contact authors for color figures.

6 Conclusion

This paper has looked at how one can infer the multi-trader heterogeneity in a financial market. The market itself could be an artificial one (i.e., simulated, like the so-called Minority Game), or a real one – the technique is essentially the same. The method we presented provides a significant extension of previous work [6]. When coupled with an underlying market model that better suits the time-series under analysis, these techniques could provide useful insight into the composition of multi-trader populations across a wide range of markets. We have also provided a framework for dealing with markets containing a very large number of active agents. Together, these ideas can yield superior prediction estimates of a real financial time-series.

References

1. Challet D, Marsili M, Zhang YC (2005) Minority Games, Oxford University Press, Oxford.
2. Jefferies P, Johnson NF (2002) Designing agent-based market models; e-print cond-mat/0207523 at xxx.lanl.gov
3. Johnson NF, Jefferies P, Hui PM (2003) Financial Market Complexity, Oxford University Press, Oxford.
4. Johnson NF, Lamper D, Jefferies P, Hart ML, Howison SD (2001) Application of multi-agent games to the prediction of financial time-series, Physica A 299: 222–227
5. Andersen JV, Sornette D (2005) A mechanism for pockets of predictability in complex adaptive systems, Europhysics Letters 70:697–703
6. Gupta N, Hauser R, Johnson NF (2005) Using artificial market models to forecast financial time-series. In: Workshop on Economic Heterogeneous Interacting Agents 2005, e-print physics/0506134 at xxx.lanl.gov.
7. Bar-Shalom Y, Li XR, Kirubarajan T (2001) Estimation with Applications to Tracking and Navigation. John Wiley and Sons, Inc.
8. Chiang YT, Wang LS, Chang FR, Peng HM (2002) Constrained filtering method for attitude determination using gps and gyro, IEE Proceedings - Radar, Sonar, and Navigation, 149:258–264
9. Wang LS, Chiang YT, Chang FR (2002) Filtering method for nonlinear systems with constraints, IEE Proceedings - Control Theory and Applications 149:525–531
10. Simon D, Simon DL (2003) Kalman filtering with inequality constraints for turbofan engine health estimation, Technical Report A491414, National Aeronautics and Space Administration, John H. Glenn Research Center at Lewis Field
11. Maybeck PS (1982) Stochastic Models, Estimation and Control. Volume 2. Academic Press, Inc.
12. Friedland B (1969) Treatment of bias in recursive filtering, IEEE Transactions on Automatic Control 14:359–367
13. Casella G, Berger RL (2002) Statistical Inference, Thomson Learning

Part II

Business and Trade Networks

Dynamical Structure of Behavioral Similarities of the Market Participants in the Foreign Exchange Market

Aki-Hiro Sato and Kohei Shintani

Department of Applied Mathematics and Physics, Graduate School of Informatics, Kyoto University, Kyoto 606-8501, Japan.
aki@i.kyoto-u.ac.jp

1 Introduction

The financial markets started to be computerized due to development and spread of the Information and Communication Technology (ICT) in early 1990s. As the result rapid development and spread of electrical trading systems occurred all over the world. Moreover advance of processing speed of computers and capacity of storages leads to accumulation of activity records of market participants, high frequency financial data. By utilizing the high frequency financial data one can observe behavior of the market participants with high resolutions and analyze a large amount of data enough to quantify them in the statistically significant.

Econophysics [1–6] emerged during the same period as spreading of ICT. It seems not to be a chance but to be a necessity that these occurrence were coincident. Since ICT makes uncertainty of human activities due to lack of information decreasing, human activities which were not observed once become detectable. Physics needs high-accuracy data in order to establish theoretical concept. At the beginning of 1990s supply of data about human activities and demand for theory of human met together. Presently we seem to stand at the next stage of scientific understandings of human activities.

This article focuses on the tick frequencies, which represents quotation activity of the market participants, and quantifies their similarities among currency pairs which the traders exchange in the foreign exchange market. This analysis provides insights about both perception and action of the market participants in the foreign exchange market.

2 Foreign Exchange Market

The foreign exchange market is the largest financial market all over the world. Its turnover in 2004 was estimated as 1.9 trillion USD. The foreign exchange market means the networks among financial institutions to exchange currencies. The market participants mainly belong to international banks, regional banks, and other financial institutions around the world. The currencies are mainly traded in three areas; the Asia including Japan and Singapore, the Europe including German and UK, and the America including US. The foreign exchange market is open for 24 hours on weekdays, and is close on Saturdays and Sundays. The market participants in the foreign exchange market trade almost through the electrical broking systems (the EBS and the Reuters3000).

In this article the data provided by CQG Inc. is investigated [8]. The data contains ask/bid rates and time stamps with 1 [min] resolution. The source of this data is the EBS of which terminal computers are used by more than 2,000 traders in 800 dealing rooms.

The market participants in the foreign exchange market adopt the two-way quotation, where both buyers and sellers quote both buying and selling rates at the same time. Therefore one cannot estimate excess demand from the quotation records. As shown in Table 1 the data records the almost same numbers of ask quotes and bid quotes.

Table 1. Sample of the tick data of the USD/JPY. The dates, rates, and indicators to show ask quotes (A) and bid quotes (B) are recorded.

date	time	rate	A(Ask)/B(Bid)
20000904	1708	10589	B
20000904	1708	10594	A
20000904	1708	10588	B
20000904	1708	10593	A
20000904	1710	10583	B
20000904	1710	10588	A

3 Tick Frequencies

Since all the action of the market participants in the foreign exchange market are only quotation through terminal computers connected to server computers, it is thought that quotation behavior represents minds of the market participants. The tick frequency of the ith currency pair is defined as

$$A_i(k) = \frac{1}{\Delta t} N(k\Delta t; (k+1)\Delta t), \quad (k = 0, 1, \ldots), \tag{1}$$

where $N(t_1; t_2)$ denotes the number of the ask quotation in $[t_1, t_2)$ and Δt is a sampling interval and $\Delta t = 1$ [min] throughout this analysis. If $A_i(k)$ is high/low then the ith currency pair is actively traded.

4 Motivation

To avoid risks resulting from "explosive information" in recent years computer-based trading algorithms called "algorithmic trading" have been studied. Presently many automatic trading machines work in real financial markets. Specifically in the foreign exchange market about 10% market participants introduced the automatic trading interface in 2006 [9]. More becoming common more quickly trading can be conducted. Moreover the automatic trading systems have an advantage to human traders from the viewpoint of accuracy, speed, and cost-efficiency. If automatic trading systems spread in the financial market then human beings may not be able to understand what occurs in financial markets in real-time. In order to understand the financial market and make them stabilize we propose to construct systems to monitor whole pictures of the financial markets. It is analogous to a control system used when airplanes are operated.

In this article quantitative methods based on the tick data are introduced and intuitive interpretation of results are shown. Specifically in order to investigate complex systems explosive information must be handled with information condensation. One faces two kinds of difficulties in order to deal with explosive information; (1) computational complexity, and (2) human recognition. To compute the behavioral frequency provides us reduction of massive information. Furthermore behavioral frequencies are useful to understand the states of the market participants in the financial markets from viewpoint of behavioral sciences. The market prices are determined based on minds of the market participants but minds are unobservable. Human behavior is observable and may be related to the minds.

Therefore observation and quantification of behavioral frequencies are expected to provide information about minds of the market participants indirectly. We will present appropriateness and availability of measuring behavioral frequencies through model analysis and empirical analysis.

5 Quantification of Behavioral Similarities

In order to quantify states of the financial markets we focus on the similarity between tick frequencies. If the two tick frequencies similarly vary then they are expected to have similar generating mechanism (behavior of the market participants). Therefore similarity structure of the tick frequency is one of characteristics to understand the states of the financial markets.

Several methods for quantifying similarity (Pearson's correlation is regarded as classical method, and phase only correlation, instantaneous phase, and the Kullback-Leibler spectral distance) are introduced. In the following subsections definitions of similarity measure are addressed.

5.1 Pearson's Product-Moment Correlation

Pearson's product-moment correlation r_{ij} between two tick frequencies $\{A_i(k)\}_{k=0}^{N-1}$, and $\{A_j(k)\}_{k=0}^{N-1}$ of currency pairs i, and j is defined as

$$r_{ij} = \frac{\text{cov}(A_i, A_j)}{\sqrt{\text{var}(A_i)}\sqrt{\text{var}(A_j)}}, \tag{2}$$

where $\langle A_i \rangle = \frac{1}{N}\sum_{k=0}^{N-1} A_i(k)$, $\text{cov}(A_i, A_j) = \frac{1}{N}\sum_{k=0}^{N-1}(A_i(k) - \langle A_i \rangle)(A_j(k) - \langle A_j \rangle)$, and $\text{var}(A_i) = \frac{1}{N}\sum_{k=0}^{N-1}(A_i(k) - \langle A_i \rangle)^2$. $|r_{ij}| \leq 1$. $r_{ij} = 1$ means that two signals have complete co-occurring relation. $r_{ij} = 0$ means that two signals have decorrelation.

5.2 Phase only Correlation Coefficient

Phase only correlation is calculated as similarity between two phase spectra. For data set $\{X(k)\}_{k=0}^{N-1}$ discrete Fourier transform and discrete inverse Fourier transform are defined as $\mathfrak{F}[X(k)](f_k) = \tilde{X}(f_k) = \sum_{n=0}^{N-1} X(k) \times \exp\left(-2\pi \mathrm{i} k \frac{n}{N}\right)$ and $\mathfrak{F}^{-1}[X(f_k)](k) = \frac{1}{N}\sum_{n=0}^{N-1} \tilde{X}(f_k) \exp\left(2\pi \mathrm{i} k \frac{n}{N}\right)$, where $f_n = \frac{n}{N\Delta t}$ $(n = 0, \cdots, N/2)$ denotes nth frequency. The Nyquist critical frequency is $f_c = 1/(2\Delta t) = 1/2$ [1/min] throughout this analysis.

The discrete Fourier transform of the tick frequency $\{A_i(k)\}_{k=0}^{N-1}$ and $\{A_j(k)\}_{k=0}^{N-1}$ are represented as $\{\tilde{A}_i(f_n)\}_{n=0}^{N-1}$ and $\{\tilde{A}_j(f_n)\}_{n=0}^{N-1}$, respectively. Then the phase only correlation POC_{ij} is defined as

$$POC_{ij}(k) = \mathfrak{F}^{-1}\left[\frac{\tilde{A}_i(f_n)\overline{\tilde{A}_j(f_n)}}{|\tilde{A}_i(f_n)||\tilde{A}_j(f_n)|}\right](k) \quad (k = 0, \cdots, N-1), \tag{3}$$

where $\overline{X}$ denotes the conjugate complex number of X. Afterward $POC_{ij} = POC_{ij}(k)|_{k=0}$ is called phase only correlation coefficient. $POC_{ij} = 1$ means that phase spectra of ith signal and jth signal are same. $POC_{ij} = 0$ means that they are completely dissimilar.

5.3 Instantaneous Phase

Similarity between instantaneous phases, defined by analytic signal, which was advocated by Gabor is considered. To utilize the similarity between instantaneous phases is useful to characterize the similarity of two signal from viewpoint of synchronization.

The analytic signal $\{\hat{X}(k)\}_{k=0}^{N-1}$ of original signal $\{X(k)\}_{k=0}^{N-1}$ is defined as

$$\hat{X}(k) = X(k) + \mathrm{i}[-\mathrm{i} \times \text{sgn}(f_n)\tilde{X}(f_n)](t), \tag{4}$$

$$\text{sgn}(f) = \begin{cases} 1 & (f > 0) \\ -1 & (f < 0) \\ 0 & (\text{otherwise}) \end{cases}, \tag{5}$$

where $\hat{X}(f_n)$ denotes the discrete Fourier transform of $\{X(k)\}_{k=0}^{N-1}$. $\hat{X}(k)$ $(k = 0, \cdots, N-1)$ can be separated into phase and amplitude, and written as $\hat{x}(k) = B_k \exp(\phi(k))$. The phase components $\phi(k)$ $(k = 0, \cdots, N-1)$ are called instantaneous phase at k of the analytic signal $\{\hat{X}(k)\}_{k=0}^{N-1}$ of original signal $\{X(k)\}_{k=0}^{N-1}$.

By using phase components the similarity between two tick frequencies is quantified as

$$PS_{ij} = \left\{ \frac{1}{N} \sum_{k=0}^{N-1} \sin\left(\phi_i(k) - \phi_j(k)\right) \right\}^2 + \left\{ \frac{1}{N} \sum_{k=0}^{N-1} \cos\left(\phi_i(k) - \phi_j(k)\right) \right\}^2 . \quad (6)$$

If $PS_{ij} = 1$ then the two signals are completely synchronized. On the other hand $PS_{ij} = 0$ means they are desynchronized.

5.4 Kullback-Leibler Spectral Distance

The Kullback-Leibler spectral distance quantifies the similarity between power spectra of two tick frequencies. If the power spectra are similar then their generating processes seem to be similar. It is an application of the relative entropy, which is widely used in information theory, information geometry, phonetic analysis, and so on, to a normalized power spectrum. Using the discrete Fourier transform $\tilde{X}(f_k)_{k=0}^{N/2}$ of $\{X(k)\}_{k=0}^{N-1}$ power spectrum $S(f_n)$ at nth frequency f_n is estimated as periodgram estimator. Using the normalized power spectrum, defined as $s(f_k) = S(f_k)/\sum_{k=1}^{N/2} S(f_k)$, the Kullback-Leibler spectral distance KL_{ij} for $\{s_i(f_k)\}_{k=1}^{N/2}$ and $\{s_j(f_k)\}_{k=1}^{N/2}$ is defined as

$$KL_{ij} = \sum_{k=1}^{N/2} s_i(f_k) \log\left[\frac{s_i(f_k)}{s_j(f_k)}\right], \quad (7)$$

Note that in this definition a direct current component is ignored. KL_{ij} does not have symmetric property, and does not fill the axiom of the distance. Here symmetrical Kullback-Leibler spectral distance (J-divergence) SKL_{ij} which is defined as

$$SKL_{ij} = \sum_{k=1}^{N-1} [s_i(f_k) - s_j(f_k)] \log\left[\frac{s_i(f_k)}{s_j(f_k)}\right], \quad (8)$$

is used. SKL_{ij} has both symmetric and non-negative properties; $SKL_{ij} = SKL_{ji}$ and $SKL_{ij} \geq 0$. $SKL_{ij} = 0$ if and only if $s_i(f_n) = s_j(f_n)$ almost everywhere. Afterward for correspondence with other methods the transformation $K_{ij} = \exp[-SKL_{ij}]$ to the SKL is adopted. If $K_{ij} = 1$ then power spectra are completely identical. On the other hand $K_{ij} = 0$ represents that they are completely different from each other.

6 Agent-based Model

In this section an agent-based model composed of double-threshold agents is introduced [10]. Pseudo tick frequencies are produced by computer simulation, and relations between similarities between two tick frequencies and agent parameters are investigated.

6.1 Agent Behavior

The i-th market participant ($i = 1, \cdots, N$) perceives information $x_i(t)$ at time t ($t = 1, 2, \cdots$). The information is assumed to be quantified as a scalar value. The more it is positive/negative, the more he/she wants to buy/sell financial commodities actively.

The i-th market participant interprets information $x_i(t)$. The interpretation of the same information depends on the market participants, and can change due to their experience. Therefore the interpretation $\Xi_i(x_i(t), t)$ is a function of time t and the information $x_i(t)$ and the function form depends on the market participants. One of the most simplest function form is a linear function of $x_i(t)$ and a noise. Here the uncertainty of the interpretation is assumed to be described as an additive Gaussian noise:

$$\Xi_i\left(x_i(t), t\right) = x_i(t) + \xi_i(t), \quad \xi_i(t) : N\left(0, \sigma^2\right). \tag{9}$$

The interpretation drives feeling about the information. It is assumed that the feeling has one-to-one relation to the interpretation. Namely the feeling can be identical to the interpretation.

The market participants judge whether they can accept their feeling or not. Quantitatively the feeling about the feeling of the i-th market participant for the j-th financial commodity $\Phi_{ij}(t)$ may be described as a function of both the feeling $x_i(t) + \xi_i(t)$ and a multiplicative factor $a_{ij}(t)$,

$$\Phi_{ij}(t) = a_{ij}(t)\left(x_i(t) + \xi_i(t)\right). \tag{10}$$

If $a_{ij}(t)$ is positive/negative then the feeling about the feeling supports/refutes the feeling.

Action of the market participants are restricted as "buying", "selling", and "waiting". The i-th market participant determines his/her investment attitude for the j-th currency pair at time t, $y_{ij}(t)$, based on the feeling about the feeling $\Phi_{ij}(t)$.

According to the Granovetter model [11] if the inner state of agents excesses a threshold value the agents decide to take an action nonlinearly. Applying this mechanism to the market participants behavior the investment attitude can be modeled. In order to divide three action the two thresholds are needed at least. It is natural that action (buying/selling) can appear if $\Phi_{ij}(t)$ is a large positive/negative value. Namely it is assumed that the i-th market participant has two thresholds to decide buying and selling, $\theta_{ij}^B(t)$ and $\theta_{ij}^S(t)$ $(\theta_{ij}^B(t) > \theta_{ij}^S(t))$ and that the action $y_{ij}(t)$ can be described as

$$
y_{ij}(t) = \begin{cases} 1 & (\Phi_{ij}(t) \geq \theta_{ij}(t)^B) \\ 0 & (\theta_{ij}(t)^S < \Phi_{ij}(t) < \theta_{ij}(t)^B) \\ -1 & (\Phi_{ij}(t) \leq \theta_{ij}^S(t)) \end{cases}, \tag{11}
$$

where "buying", "selling", and "waiting" are coded as 1, −1, and 0, respectively. The excess demand and the tick frequency of the j-th financial commodity are defined as $D_j(t) = N^{-1}\sum_{i=1}^N y_{ij}(t)$, and $A_j(t) = \frac{1}{\Delta t}\sum_{i=1}^N |y_{ij}(t)|$, respectively. Logarithmic returns $r_j(t) = \log P_j(t) - \log P_j(t - \Delta t)$ ($P_j(t)$ denotes the market price of the j-th financial commodity at time t) can be approximated as a product of the excess demand and the liquidity constant: $r_j(t) = \gamma D_j(t)$ $(\gamma > 0)$.

The information $x_i(t)$ perceived by the i-th market participant at time t can be divided into endogenous information (weighted moving average of returns) and exogenous information (news):

$$
x_i(t) = \sum_{k=1}^M c_{ik}\left(\theta_{ik}^B(t), \theta_{ik}^S(t)\right) \frac{1}{T_{ik}(t)} \sum_{\tau=1}^{T_{ik}(t)} r_j(t - \Delta t) + s_i(t), \tag{12}
$$

where $T_{ij}(t)$ represents the memory length of the i-th market participant for the j-th financial commodity at time t. $c_{ij}(\theta_{ij}^B(t), \theta_{ij}^S(t))$ is the attention of the i-th market participant to the j-th financial commodity. The more absolute values of the thresholds are, the less market participants become active. Namely the market participants do not focus on the financial commodities if the absolute values of the thresholds are high. Therefore $c_{ij}(x, y)$ is monotonically decreasing function of $|x|$ and $|y|$. For simplicity $c_{ij}(x, y)$ is assumed as $c_{ij}(x, y) = 1/(x^2 + y^2)$. Since the exogenous information $s_i(t)$ is unpredictable it seems to be reasonable to describe it as a random variable which obeys the normal distribution, $s_i(t) : N(0, \sigma_s^2)$.

6.2 Numerical Simulation

For convenience of numerical simulation it is assumed that $a_{ij}(t)$ can be divided into the common component of the market participants and the diversification of the market participants and that they can be described as random variables which obey the identical and independent normal distribution, $a_{ij}(t) = a(t) + a_{ij}'(t)$, where $a_{ij}(t) : N(0, \sigma_a^2)$, and $a_{ij}'(t) : N(0, \sigma_{a'}^2)$. Of course this assumption can be weakened, for example, colored noises can be applied to $a_{ij}(t)$.

For $M = 2$, $N = 500$, $T_{ij}(t) = 5$, $\Lambda^B = 0.23$, $\Lambda^S = -0.23$, $\sigma_a = 0.315$, $\sigma_{a'} = 0.0315$, $\sigma_\xi = \sigma_s = 0.2$ and $\gamma = 0.1$ numerical simulation is performed. $\theta_{ij}^S(t)$ and $\theta_{ij}^B(t)$ are constant in time.

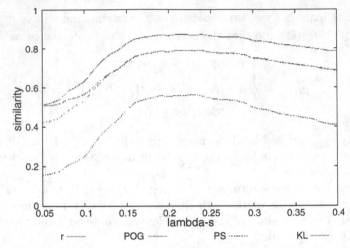

Fig. 1. The relation between similarities between two tick frequencies and Λ_2^S at $\Lambda_1^S = 0.23$. r represents Peason's product-moment correlation, POC phase only correlation, PS similarity between instantaneous phases, and KL the symmetric Kullback-Leibler distance, respectively. The similarities have a peak at $\Lambda_1^S = \Lambda_2^S$.

Figure 1 shows the relation between Λ_j^S and the similarity. It is found that the similarity for all the methods takes the maximum value at $\Lambda_2 = 0.23$, which is the same value as $\Lambda_1 = 0.23$. Alternatively the difference between Λ_1^S and Λ_2^S becomes bigger the similarity becomes lower. The difference between thresholds largely influences the similarity.

Since four kinds of similarity measures provide almost same results, similarity of behavioral parameters of the market participants is related to similarity of behavioral frequencies despite similarity measures. It is concluded that the behavioral frequencies tend to be similar if the behavioral parameters are similar. Contrarily when the behavioral frequencies are similar one may infer that the behavioral parameters of the market participants are similar. At least when the behavioral frequencies are different from each other the behavioral parameters of market participants trading those financial commodities are different from each other.

7 Empirical Results

In this section brief comments for empirical results of actual tick data of the foreign exchange market are addressed. The similarities among currency pairs are calculated by using four similarity measures from three aspects; (1) macroscopic time scale, (2) microscopic time scale, and (3) networks of currency pairs. Three facts are clarified through empirical analysis; (a) relations among currency pairs have tendency to be more similar than before

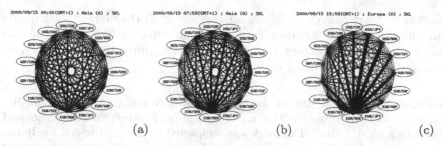

(a)   (b)   (c)

Fig. 2. The symmetric Kullback-Leibler spectral distances at (**a**) Asian time zone, (**b**) European time zone, and (**c**) American time zone among 15 kinds of currency pairs (EUR/CHF, EUR/GBP, EUR/JPY, EUR/NOK, EUR/SEK, EUR/USD, NZD/USD, USD/CAD, USD/CHF, USD/JPY,USD/NOK, and USD/SEK) on 15th September 2000. *Nodes* represent currency pairs and *thick/thin lines* similarity/dissimilarity of two connected currency pairs. The similarity structure depends on the time zone.

from viewpoint of macroscopic time scale, (b) similarity structure varies in time from microscopic time scale, and (c) similarity structure has patterns in time (See Fig. 2).

8 Conclusion

The dynamics of the foreign exchange market was analyzed through similarity of the tick frequency, which is the trace of the action by market participants. By utilizing four kinds of similarity measures (Pearson's correlation, phase only correlation coefficient, similarity of instantaneous phase, and the Kullback-Leibler spectral distance) similarity structure of the tick frequencies among several currency pairs was computed. As the results it was found that similarity structure of the tick frequencies dynamically varies with rotation of the earth and that four kinds of methods have the same tendency. However the similarity between the instantaneous phases was slightly different from others. In order to confirm appropriateness of four similarity measures the agent-based model of a financial market in which N market participants exchange M kinds of financial commodities was introduced. From numerical simulation it was confirmed that each similarity measure becomes high when the behavioral parameters to determine both perception and action for the market participants trading those financial commodities are similar. Hence the similarity of the tick frequency is associated with behavioral parameters of the market participants.

Computerization of financial markets may have brought potential risks which damage the world economy; (1) operational risks, (2) market risks, and (3) systemic risks. In order to avoid these risks, constructive utilization of the tick data should be sophisticated. We propose that a real-time monitoring system of the financial markets should be constructed from physical point of view in order to avert emergency of market crashes. If all the market participants can understand their situation then mass panic due to lack of information may be averted. We expect that this work contributes to both safety and stabilization of the financial markets.

Acknowledgement. This work was supported by a Grant-in-Aid for Scientific Research (#17760067) from the Ministry of Education, Culture, Sports, Science and Technology (A.-H. Sato). This work was supported by a Grant-in-Aid for Initiatives for Attractive Education in Graduate Schools from the Ministry of Education, Culture, Sports, Science and Technology (K. Shintani).

References

1. Sato AH, Takayasu H (1998) Dynamical models of stock market exchanges: from microscopic determinism to macroscopic randomness, Physica A 250: 231–252.
2. Lux T, Marchesi M (1999) Scaling and criticality in a stochastic multi-agent model of a financial market, Nature 397: 498–500.
3. Mantegna RN, Stanley HE (2000) An Introduction to Econophysics – Correlations and Complexity in Finance, Cambridge University Press, Cambridge.
4. Dacorogna MM, Gençay R, Müller U, Olsen RB, Pictet OV (2000) An introduction to high-frequency finance, Academic Press, San Diego.
5. Takayasu H (2006) Ed. Practical Fruits of Econophysics, Springer-Verlag (Tokyo).
6. Sato AH (2006) Frequency analysis of tick quotes on foreign currency markets and the double-threshold agent model, Physica A 369: 753–764
7. Sato AH (2006) Characteristic time scales of tick quotes on foreign currency markets: empirical study and agent-based model, European Physical Journal B, 50: 137–140.
8. The data are provided by CQG International Ltd.
9. The information is available at EBS homepage: http://www.ebs.com.
10. Sato AH (2006) Frequency analysis of tick quotes on the foreign exchange market and agent-based modeling: A spectral distance approach, http://arxiv.org/abs/physics/0607273.
11. Granovetter M (1978), Threshold models of collective behavior, The American Journal of Sociology, 83: 1420–1443.

Weighted Networks at the Polish Market

A.M. Chmiel, J. Sienkiewicz, K. Suchecki, and J.A. Hołyst

Faculty of Physics and Center of Excellence for Complex Systems Research, Warsaw University of Technology, Koszykowa 75, PL 00-662 Warsaw, Poland
jholyst@if.pw.edu.pl

1 Introduction

During the last few years various models of networks [1,2] have become a powerful tool for analysis of complex systems in such distant fields as Internet [3], biology [4], social groups [5], ecology [6] and public transport [7]. Modeling behavior of economical agents is a challenging issue that has also been studied from a network point of view. The examples of such studies are models of financial networks [8], supply chains [9,10], production networks [11], investment networks [12] or collective bank bankrupcies [13,14]. Relations between different companies have been already analyzed using several methods: as networks of shareholders [15], networks of correlations between stock prices [16] or networks of board directors [17]. In several cases scaling laws for network characteristics have been observed.

In the present study we consider relations between companies in Poland taking into account common branches they belong to. It is clear that companies belonging to the same branch compete for similar customers, so the market induces connections between them. On the other hand two branches can be related by companies acting in both of them. To remove weak, accidental links we shall use a concept of threshold filtering for weighted networks where a link weight corresponds to a number of existing connections (common companies or branches) between a pair of nodes.

2 Bipartite Graph of Companies and Branches

We have used the commercial database "Baza Kompass Polskie Firmy B2B" [18] from September 2005. It contains information concerning over 50 000 large and medium size Polish companies belonging to one or more of 2150 different branches. A bipartite graph of companies and branches has been constructed as at Fig. 1.

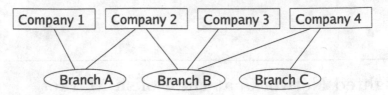

Fig. 1. Bipartite graph of companies and trades.

In the bipartite graph we have two kinds of objects: branches $A = 1, 2, 3 \ldots N_b$ and companies $i = 1, 2, 3 \ldots \ldots N_f$, where N_b – total number of branches and N_f – total number of companies. Let us define a *branch capacity* $|Z(A)|$ as the cardinality of set of companies belonging to the branch A. At Fig. 1 the branch A has the capacity $|Z(A)| = 2$ while $|Z(B)| = 3$ and $|Z(C)| = 1$. The largest capacity of a branch in our database was 2486 (construction executives), the second largest was 2334 (building materials).

Let $B(i)$ be a set of branches a given company i belongs to. We define a *company diversity* as $|B(i)|$. An average company diversity μ is given as

$$\mu = \frac{1}{N_f} \sum_{i=1}^{i=N_f} |B(i)|. \tag{1}$$

For our data set we have $\mu = 5.99$.
Similarly an average branch capacity ν is given as

$$\nu = \frac{1}{N_b} \sum_{A=1}^{A=N_b} |Z(A)|, \tag{2}$$

and we have $\nu = 134$.
It is obvious that the following relation is fulfilled for our bipartite graph:

$$\mu N_f = \nu N_b. \tag{3}$$

3 Companies and Branches Networks

The bipartite graph from Fig. 1 has been transformed to create a *companies network*, where nodes are companies and a link means that two connected companies belong to at least one common branch. If we used the example from Fig. 1 we would obtain a companies network presented at Fig. 2.

We have excluded from our dataset all items that correspond to communities (local administration) and as a result we got $N_f = 48158$ companies, belonging to a single connected cluster. Similarly a *branch network* has been constructed where nodes are branches and an edge represents connection if at least one company belongs to both branches. In our database we have $N_b = 2150$ different branches.

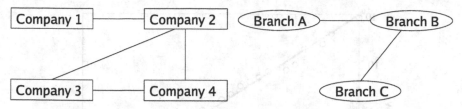

Fig. 2. Companies network on the *left*, branches network on the *right*.

4 Weight, Weight Distribution and Networks with Cutoffs

We have considered link-weighted networks. In the branches network the link weight means a number of companies that are active in the same pair of branches and it is formally a cardinality of a common part of sets $Z(A)$ and $Z(B)$, where $Z(A)$ is a set of companies belonging to the branch A and $Z(B)$ is a set of companies belonging to the branch B:

$$w_{AB} = |Z(A) \cap Z(B)|. \tag{4}$$

Let us define a function f_k^A which is equal to one if a company k belongs to the branch A, otherwise it is zero:

$$f_k^A = \left\{ \begin{array}{l} 1, k \in A \\ 0, k \notin A \end{array} \right\}. \tag{5}$$

Using the function f_k^A the weight can be written as:

$$w_{AB} = \sum_{k=1}^{N_F} f_k^A \, f_k^B. \tag{6}$$

The weight distribution $p(w)$, meaning the probability p to find a link with a given weight w, is presented at Fig. 3. The distribution is well approximated by a power function

$$p(w) \sim w^{-\gamma}, \tag{7}$$

where the exponent $\gamma = 2.46 \pm 0.07$. One can notice the existence of edges with large weights. The maximum weight value is $w_{max} = 764$, and the average weight

$$\langle w \rangle = \sum_{w_{min}}^{w_{max}} w p(w), \tag{8}$$

equals $\langle w \rangle = 4.67$.

Using cutoffs for link weights we have constructed networks with different levels of filtering. In such networks nodes are connected only when their edge weight is no less than an assumed cutoff parameter w_o. In Table 1 we present

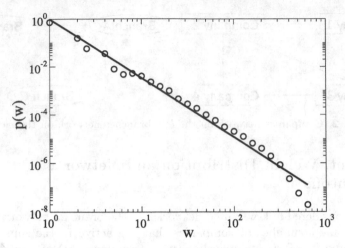

Fig. 3. Weight distribution in branches network.

Table 1. Data for branches networks: w_o is the value of selected weight cutoff, N is the number of vertices with nonzero degrees, E is the number of links, k_{max} is the maximum node degree, $\langle k \rangle$ is the average node degree, C is the clustering coefficient.

w_o	N	E	k_{max}	$\langle k \rangle$	C
1	2150	389542	1716	362	0.530
2	2109	212055	1381	201	0.565
3	2053	136036	1127	132.	0.568
4	2007	100917	952	100	0.575
5	1948	80358	802	82	0.589
6	1904	66353	655	69	0.592
7	1858	56565	569	60	0.596
8	1819	49193	519	54	0.597
9	1786	43469	477	48	0.599
10	1748	38924	450	44	0.600
12	1666	32167	394	38	0.615
14	1611	26088	325	32	0.605
16	1545	21762	288	28	0.606
18	1490	18451	259	24	0.603
20	1424	15872	226	22	0.604
30	1188	8989	162	15	0.585
40	996	6036	131	12	0.587
50	857	4379	111	10	0.572
60	752	3303	85	8	0.551
70	666	2638	65	7	0.524
80	575	2143	55	7	0.532
90	512	1808	49	7	0.538
100	464	1543	41	6	0.546
150	306	750	26	4	0.493

the main parameters of branches networks with changing w_o. The clustering coefficient has been calculated by the definition:

$$C = \frac{1}{N} \sum_{i=1}^{N} \frac{2E_i}{k_i(k_i - 1)} \qquad (9)$$

where E_i is the number of existing connections between neighbours of node i and k_i is the degree of vertex i. The values of clustering coefficient is very high in comparison to random networks, and it remains almost at a constant level for the presented values of cutoff parameter.

A weight in the companies network is defined in a similar way as in the branches networks, it is the number of common branches for two companies – formally it is equal to the cardinality of a common part of sets $B(i)$ and $B(j)$, where $B(i)$ is a set of branches the company i belongs to, $B(j)$ is a set of branches the company j belongs to:

$$w_{ij} = |B(i) \cap B(j)|. \qquad (10)$$

Using the function f_k^A the weight can be written as

$$w_{ij} = \sum_{A=1}^{N_b} f_i^A \, f_j^A. \qquad (11)$$

Table 2. Data for companies networks: w_o is the selected cutoff, N is the number of nodes with nonzero degrees, E is the number of links, k_{max} is the maximum node degree, $\langle k \rangle$ is the average node degree, C is the clustering coefficient.

w_o	N	E	k_{max}	$\langle k \rangle$	C
1	48158	39073685	16448	1622	0.652
2	39077	9932790	8366	508	0.689
3	31150	3928954	4842	252	0.714
4	24212	1895373	3103	156	0.717
5	18566	1024448	2059	110	0.713
6	14116	622662	1412	88	0.710
7	10796	404844	1012	74	0.700
8	8347	266013	724	63	0.701
9	6527	180696	566	55	0.699
10	5197	124079	443	47	0.699
11	4268	94531	382	44	0.704
12	3400	68648	345	40	0.693
13	2866	54258	305	37	0.691
14	2277	36461	277	32	0.663
15	1903	28844	249	30	0.673
16	1627	23063	231	28	0.678
17	1397	18352	212	26	0.667
18	1196	14480	191	24	0.680
19	1003	11230	171	22	0.680
20	883	8907	159	20	0.676

The maximum value of observed weights $w_{max} = 207$ in this network is smaller than in the branches network while the average value equals $\langle w \rangle = 1.48$. The weight distribution in this case does not follow a power law and in a limited range it shows an exponential behavior.

Similarly to the branches networks we have introduced cutoffs in companies network. The dependence of selected network parameters on cutoff threshold is shown in Table 2. The behaviour of clustering coefficient resembles the one observed in the branches networks.

At Fig. 4 we present average degrees of nodes and maximum degrees as functions of the cutoff parameter w_o. We have observed a power law scaling

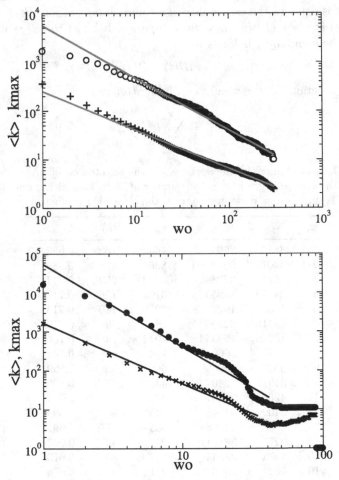

Fig. 4. Dependence of $\langle k \rangle$ and k_{max} on cutoff parameter w_o for branches networks (*top*) and companies networks (*bottom*).

$$\langle k \rangle \sim w_o^{-\beta}, \tag{12}$$

$$k_{max} \sim w_o^{-\alpha}, \tag{13}$$

where for branches networks $\alpha_b = 1.069 \pm 0.008$ and $\beta_b = 0.792 \pm 0.005$ while for companies networks $\alpha_f = 2.13 \pm 0.07$ and $\beta_f = 1.55 \pm 0.04$.

5 Degree Distribution

We have analyzed the degree distribution for networks with different cutoff parameters. At Fig. 5 we present the degree distributions for companies networks for different values of w_o. The distributions change qualitatively with increasing w_o from a nonmonotonic function with an exponential tail (for $w_o = 1$) to a power law with exponent γ (for $w_o > 6$).

At the Fig. 6 we present a degree distribution for branches networks. For $w_o = 1$ we observe a high diversity of node degrees – vertices with large values of k occur almost as frequent as vertices with a small k. For a properly chosen cutoff values the degree distributions are described by power laws. For $w_o = 4$ we see two regions of scaling with different exponents γ_1 and γ_2 while a transition point between both scaling regimes appears at $k \approx 100$. Branches belonging to the first regime of scaling are more specific, for example "production of protective clothing", "poultry farmer", "gemstone" and branches on the right are more general like "import and export general", "network of supermarkets". We suppose that the mechanism of this behaviour is similar to

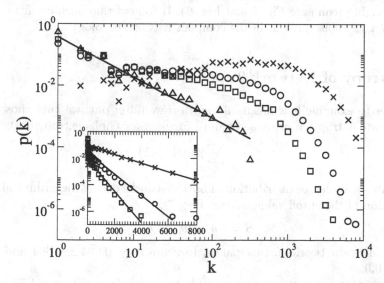

Fig. 5. Degree distributions for companies networks for different values of w_o. *X-marks* are for $w_o = 1$, *circles* are for $w_o = 2$, *squares* are for $w_o = 3$ and *triangles* are for $w_o = 12$.

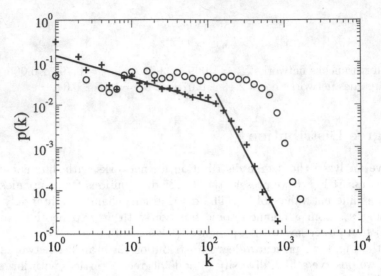

Fig. 6. Degree distribution in branches network for different values of w_o. *Circles* are for $w_o = 1$, *crosses* are for $w_o = 4$.

linguistic networks [19]. Appearance of a new general branch creates connections between existing specific branches, what causes double scaling in the end.

It is important to stress that in both networks (companies and branches) the scaling behavior for degree distribution occurs only if we use cutoffs for links weights (compare Fig. 5 and Fig. 6). It follows that such cutoffs act as filters for the noise present in the complex network topology.

6 Entropy of Network Topology

In order to examine how much information we fillter out, we have chosen to investigate entropy S, using a standard formula for Gibbs entropy (14):

$$S = - \sum_k p(k) \ln p(k). \tag{14}$$

The entropy of degree distribution in both networks decays logarithmically as a function of the cutoff value w_o (see Fig. 7):

$$S = -a \ln(w_o) + b. \tag{15}$$

For branches networks fitting parameters are $a = 0.834 \pm 0.004$ and $b = 6.51 \pm 0.02$.

The entropy in companies networks behaves similarly with $a = 1.79 \pm 0.05$ and $b = 8.49 \pm 0.15$. We have observed that decrease is faster with w_o than in the branches networks. This is due to different distributions of weight in both

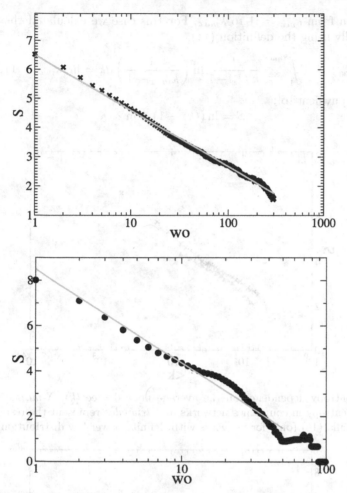

Fig. 7. Entropy dependence on cutoff parameter for branches networks on the *top* and for companies networks on the *bottom*.

networks and a smaller range of weights in the companies networks. Since they are filtered differently with w_o we have observed dependence of S on the parameter $\langle k \rangle$. This dependence is presented at the Fig. 8 along with results of formula (16) with real values of k_{max} and γ. The formula (16) is calculated for general case of power law distributions with $k_{min} = 1$.

$$S\left(\gamma, k_{max}\right) = -\ln \frac{1 - \gamma}{k_{max}^{(1-\gamma)} - 1} + \frac{\gamma k_{max}^{1-\gamma} \ln k_{max}}{k_{max}^{(1-\gamma)} - 1} - \frac{\gamma}{(1 - \gamma)} \tag{16}$$

In our system, parameters γ and k_{max} depend on each other. Both are defined by the cutoff paramter w_o. The increase of w_o results in the decrease of k_{max} and increase of γ. Let us take the simplest possible situation – a uniform distibution from $k_{min} = 1$ to k_{max}. For this case we calculated the entropy analytically using the definition (14):

$$S\left(k_{max}\right) = -\int_{1}^{k_{max}} \frac{1}{k_{max}-1} \ln\left(\frac{1}{k_{max}-1}\right) dk = \ln\left(k_{max}-1\right), \quad (17)$$

what is equivalent to :

$$S = \ln\left(\langle k \rangle - 1\right) + \ln 2. \quad (18)$$

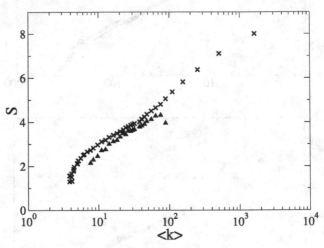

Fig. 8. Entropy dependence on the average node degree $\langle k \rangle$. *X-marks* represent measured entropy in companies networks and *triangles* represent the results of analytic formula (16) (only for networks with definite power-law distribution).

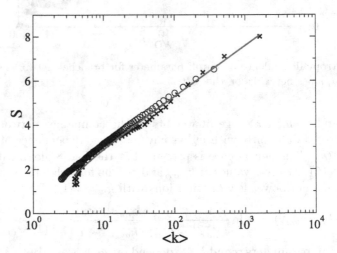

Fig. 9. Dependence of entropy on the average nodes degree. *Circles* represent branches networks and *X-marks* represent companies networks. *Line* corresponds to (18).

The value of $\langle k \rangle$ is strictly connected to the width of the distribution. Figure 9 shows that such a simplistic approach gives a very good approximation of real entropy value. We can conclude that the width of the distribution is the main source of entropy changes in our systems. The presence of two parameters (γ, k_{max}) is irrelevant as both are derived from the same parameter w_o. We decided to use the single parameter $\langle k \rangle$ since it seems to be the most significant and universal for different networks.

7 Conclusions

In this study, we have collected and analyzed data on companies in Poland. 48158 medium/large firms and 2150 branches form a bipartite graph that allows to construct weighted networks of companies and branches.

Link weights in both networks are very heterogenous and a corresponding link weight distribution in the branches network follows a power law. Removing links with weights smaller than a cutoff (threshold) w_o acts as a kind of filtering for network topology. This results in a recovery of a hidden scaling relations present in the network. The degree distribution for companies networks changes with increasing w_o from a nonmonotonic function with an exponential tail (for $w_o = 1$) to a power law (for $w_o > 6$). For a filtered ($w_o > 4$) branches network we see two regions of scaling with different exponents. Entropies of degree distributions for both networks decay logarithmically as a function of cutoff parameter and are proportional to the logarithm of the mean node degree. We have found the distribution width to be a crucial factor for entropy value.

Acknowledgement. We acknowledge a support from the EU Grant *Measuring and Modeling Complex Networks Across Domains* – MMCOMNET (Grant No. FP6-2003-NEST-Path-012999) and from Polish Ministry of Science and Higher Education (Grant No. 13/6.PR UE/2005/7), and from a special grant of Warsaw University of Technology.

References

1. Albert R, Barabási A-L (2002) Statistical mechanics of complex networks, Reviews of Modern Physics 74:47–97
2. Newman M E J (2003) The structure and function of complex networks, SIAM Review 45:167–256
3. Pastor-Satorras P, Vespignani A (2004) Evolution and Structure of the Internet: A Statistical Physics Approach, Cambridge University Press, Cambridge
4. Ravasz E, Somera A L, Mongru D A, Oltvai Z N, Barabasi A-L (2002) Hierarchical organization of modularity in metabolic networks, Science 297:1551–1555
5. Newman M E J, Park J (2003) Why social networks are different from other types of networks, Physical Review E 68:036122

6. Garlaschelli D, Caldarelli G, Pietronero L (2003) Universal scaling relations in food webs, Nature 423:165–168
7. Sienkiewicz J, Hołyst J A (2005) Statistical analysis of 22 public transport networks in Poland, Physical Review E, 72:046127
8. Caldarelli G, Battiston S, Garlaschelli D, Catanzaro M (2004) Emergence of Complexity in Financial Networks. In: Ben-Naim E, Frauenfelder H, Toroczkai Z (eds) Lecture Notes in Physics 650:399 - 423, Springer-Verlag
9. Helbing D, Lämmer S, Seidel T (2004) Physics, stability and dynamics of supply networks, Physical Review E 70:066116
10. Helbing D, Lämmer S, Witt U, Brenner T (2004) Network-induced oscillatory behavior in material flow networks and irregular business cycles, Physical Review E, 70:056118
11. Weisbuch G, Battiston S (2005) Production networks and failure avalanches, e-print physics/0507101
12. Battiston S, Rodrigues J F, Zeytinoglu H (2005) The Network of Inter-Regional Direct Investment Stocks across Europe, e-print physics/0508206
13. Aleksiejuk A, Hołyst J A (2001) A simple model of bank bankruptcies, Physica A, 299:198–204
14. Aleksiejuk A, Hołyst J A, Kossinets G (2002) Self-organized criticality in a model of collective bank bankruptcies, International Journal of Modern Physics C, 13:333–341
15. Garlaschelli G, Battiston S (2005) The scale-free topology of market investments, Physica A, 350:491–499
16. Onnela J-P, Chakraborti A, Kaski K, Kertész J, Kanto A (2003) Dynamics of market correlations: Taxonomy and portfolio analysis, Physical Review E, 68:056110
17. Battiston S, Catanzaro M (2004) Statistical properties of corporate board and director networks, European Physical Journal B 38:345–352
18. See web page of company http://www.kompass.com
19. Dorogovtsev S N, Mendes J F F (2001) Language as an evolving word web, Proceedings Royal Society of London Series B, Biological Sciences 268 (1485): 2603–2606

The International Trade Network

K. Bhattacharya[1], G. Mukherjee[1,2], and S.S. Manna[1]

[1] Satyendra Nath Bose National Centre for Basic Sciences
   Block-JD, Sector-III, Salt Lake, Kolkata-700098, India
[2] Bidhan Chandra College, Asansol 713304, Dt. Burdwan, West Bengal, India
   kunal@bose.res.in, gautamm@bose.res.in, manna@bose.res.in

Summary. Bilateral trade relationships in the international level between pairs of countries in the world give rise to the notion of the International Trade Network (ITN). This network has attracted the attention of network researchers as it serves as an excellent example of the weighted networks, the link weight being defined as a measure of the volume of trade between two countries. In this paper we analyzed the international trade data for 53 years and studied in detail the variations of different network related quantities associated with the ITN. Our observation is that the ITN has also a scale invariant structure like many other real-world networks.

From long time back different countries in the world were dependent economically on many other countries in terms of bilateral trades. A country exports its surplus products to other countries and at the same time imports a number of commodities from other countries to meet its deficit. These bilateral trades among different countries in the world have given rise to the notion of the International Trade Network. In recent years studying the structure, function and dynamics of a large number of complex networks, both in the real-world as well as through theoretical modeling have attracted intensive attention from researchers in multi-disciplinary fields [1–3] in which ITN has taken an important position in its own right [4–8]. The volume of trade between two countries may be considered as a measure of the strength of mutual economic dependence between them. In the language of graph theory this strength is known as the weight associated with the link [9]. While simple graphical representation of a network already gives much informations about its structure, it has been observed recently that in real-world networks like the Internet and the world-wide airport networks the links have widely varying weights and their distribution as well as evolution yield much insight into the dynamical processes involved in these networks [10–13].

Recently few papers have been published on the analysis of the ITN. The fractional GDP of different countries have been looked upon as the 'fitness' for the international trade. Links are then placed between a pair of nodes according to a probability distribution function of their fitnesses [5]. Also the trade imbalances between different pairs of countries, measuring the excess of export of one country to another over its import from the same country, are studied [6,8]. Using this method one can define the backbone of the ITN [8].

In the world there are different countries with different economic strengths. These countries are classified into three different categories. According to the World Bank classification of different countries in July 2005 based on gross national income (GNI) per capita as mentioned in the human development reports of 2003 [14] high income countries have GNI/capita at least $9,386, middle income countries have GNI/capita in between $9,386 and $766 where as low income countries have GNI/capita less than $766.

In a recent paper [12] we have studied the ITN as an example of the weighted networks. Analysis of the ITN data over a period of 53 years from 1948 to 2000 available in [15] have lead to the recognition of the following universal features: the link weight i.e., volume of annual trade between two countries varies over a wide range and is characterized by a log-normal probability distribution and this distribution remains robust over the entire period of 53 years within fluctuation. Secondly, the strength of a node, which is the total volume of trade of a country in a year depends non-linearly with its Gross Domestic Product (GDP). In addition a number of crucial features observed from real-data analysis have been qualitatively reproduced in a non-conservative dynamic model of the international trade using the well known Gravity model of the economics and social sciences as the starting point [16].

The annual trade between two countries i and j is described by four different quantities $\exp_{ij}$, $\exp_{ji}$, imp_{ij} and imp_{ji} measured in units of million dollars in the data available in the website [15]. In general values of $\exp_{ij}$ and imp_{ji} should be the same yet they have been quoted differently since exports from i to j and $j's$ import from i are reported as different flows in the IMF DOT data. Although magnitudes of these quantities are approximately same in most cases they do differ in many instances due to different reporting procedures followed and different rates of duties applicable in different countries etc. [17]. Therefore between two countries i and j we denote the amount of export from i to j by w_{ij}^{exp}, the amount of import from j to i by w_{ij}^{imp} and the total trade by w_{ij} and define them as:

$$w_{ij}^{exp} = \frac{1}{2}(\exp_{ij} + \text{imp}_{ji}), w_{ij}^{imp} = \frac{1}{2}(\exp_{ji} + \text{imp}_{ij}), w_{ij} = w_{ij}^{exp} + w_{ij}^{imp}. \quad (1)$$

Using these data the International Trade Network can be constructed every year. Naturally nodes of the ITN represent different countries in the world. The export w_{ij}^{exp} is the outword flow from i to j and the import w_{ij}^{imp} is the inward flow from j to i. Therefore the ITN is in general a directed graph with two opposite flows along a link, though it has been observed that few links have only one flow. Obviously one can also ignore the direction and define an undirected link between an arbitrary pair of nodes if there exists a non-zero volume of trade in any direction between the corresponding countries. Both the number of nodes N as well as the number of links L in the annual ITN varied from one year to the other. In fact they had grown almost systematically over the years. For example, the number of nodes have increased from $N = 76$ in 1948 to 187 in 2000 (Fig. 1(a)), the number of links have increased from $L = 1494$ in 1948 to 10252 in the year 2000 (Fig. 1(b)) where as the link density $\rho(N, L) = L/[(N(N-1))/2]$ fluctuated widely but with a with a slow increasing trend around a mean value of 0.52 over this period (Fig. 1(c)).

Looking at the available data few general observations can be made: Few high income [14] countries make trades to many other countries in the world. These countries form the large degree hubs of the network. In the other limit, a large number of low income countries make economic transactions to few other countries. Moreover, a rich-club of few top rich countries actually trade among themselves a major fraction of the total volume of international trade.

A huge variation of the volume of the bilateral trade is observed starting from a fraction of a million dollar to million million dollars. There are a large number of links with very small weights and this number gradually decreases to a few links with very large weights. The tail of the distribution consists of links with very large weights corresponding to mutual trades among very few high income countries [14]. The variation of the ratio of w_{max} and W is shown in Fig. 2(a). The average weight per link had also grown almost systematically

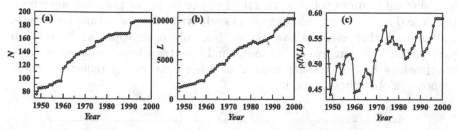

Fig. 1. Variations of the (**a**) the total number of nodes N (**b**) the total number of links L and the (**c**) link density $\rho(N, L)$ of the annual ITN over a period of 53 years from 1948 to 2000.

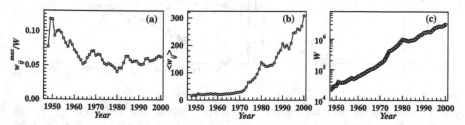

Fig. 2. Plot of different quantities over the period from 1948 to 2000: (**a**) the ratio of maximal trade w_{ij}^{max} along a link and the total volume of trade W of the ITN, (**b**) average total trade (export + import) $\langle w_{ij} \rangle$ per link and (**c**) total volume of annual world trade W.

from 15.54 M\$ in 1948 to 308.8 M\$ in 2000 (Fig. 2(b)). Again the total world
trade W had grown with years from 2.3×10^{10} dollars in 1948 to 3.2×10^{12}
dollars in 2000 (Fig. 2(c)).

The degree k of a node is the number of other countries with which this
country has trade relationships. This can be further classified by the number
of countries to which this country exports and is denoted by k_{exp} where as
k_{imp} is the number of countries from which this country imports. In general
$k_{exp} \neq k_{imp}$ but for some nodes they may be the same. The structure of
the ITN is mainly reflected in its degree distribution which has been already
studied in [5] in which a power law for the cumulative distribution $P_>(k) \sim$
$k^{1-\gamma}$ has been observed over a small range of k values with $\gamma \approx 2.6$. We have
studied the degree distributions, each averaged over ten successive ITNs, for
example, 1951–60, 1961–70, 1971–80, 1981–90 and 1991–2000. The plots are
given in Fig. 3(a). We see that indeed a small power law region appears for the
period 1991–2000 with a value of $\gamma \approx 2.74$. Such a region is completely absent
in the decade 1951–1960. In the intermediate decades similar short power law
regions are observed with larger values of γ. The average degree of a node
$\langle k \rangle$ and the maximal degree of a node k_{max} have also been studied for all
the 53 years where the size N of the ITN varied. We plot these quantities in
Fig. 3(b) using double-log scale and observe the following power law growths
as: $\langle k \rangle \sim N^{1.19}$ and $k_{max} \sim N^{1.14}$. Obviously these exponents have the upper
bound equal to unity yet they are found out to be larger than one since both
$\langle k \rangle / N$ and k_{max} / N ratios have grown slowly with time as as time progresses.
This implies that as years have passed not only more countries have taken
part in the ITN but in general individual countries have established trade
relationships with increasing number of other countries, a reflection of the
economic global liberalisation.

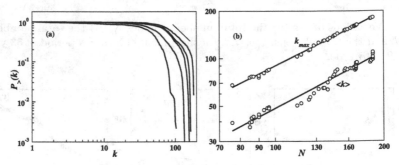

Fig. 3. (a) Cumulative degree distributions $P_>(k)$ vs. k averaged over the ten
year periods during 1951–60, 1961–70, 1971–80, 1981–90 and 1991–2000 (from left
to right). No power law variation is observed for the 1951–60 plot. For the next
decades however power laws over small regions are observed whose slopes gradually
decrease to 1.74 for the 1991–2000 plot. (b) average nodal degree $\langle k \rangle$ and the largest
degree k_{max} with the size N of the ITN.

Within a year how the ITN grows to its fully connected configuration? A flavour of this mechanism can be obtained by the following mimicry. Consider a process which starts from N nodes but with no links. Links are then inserted between pairs of nodes with a probability proportional to the weight of the link since a large weight link is more likely to be occupied than a small weight link. To do this first the link weights in the ITN have been ordered in an increasing sequence. Then the links are dropped in the descending order of the link weights starting from the maximum weight w_{max}. We have also studied the reverse procedure when links are dropped in the increasing sequence of the link weights starting from the weakest link. In the Fig. 4(a) we show the growth of the fractional size of the giant component S_g/N with the fraction f of links dropped. The plot shows that the growth rate is slower in the first case and the giant component spans the whole ITN faster than when links are dropped in the ascending order of strengths. Moreover how the single connected component is attained has been quantitatively studied by plotting $1 - S_g/N$ and f on a semi-log scale in Fig. 4(b). The intermidiate straight portions in both plots indicate exponential growths of the size of the giant component.

The annual volume of trade between a pair of countries is a measure of the strength of trade between them and is referred as the weight of the link connecting the corresponding nodes. We have studied the distribution of the total trade w_{ij} along a link in detail, without distinguishing between the exports and the imports. Therefore $\text{Prob}(w)dw$ is the probability to find a randomly selected link whose weight lies between w and $w + dw$. In general in a typical ITN, the link weights vary over a wide range. There are many many links with small weights whose number gradually decreases to a few links with large weights. In the first attempt we plot the distribution on a double logarithmic scale as shown in Fig. 5(a). Data for the six different years from 1950

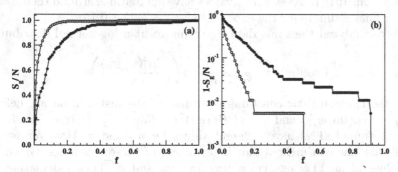

Fig. 4. (a) Fractional size S_g/N of the giant component of the ITN is plotted with the fraction f of the links that are dropped in the descending (solid) sequence from the strongest and in the ascending (opaque) sequence from the weakest weight of the links. (b) The difference $1 - S_g/N$ has been plotted on a semi-log scale which indicates an exponential approach to the fully connected network.

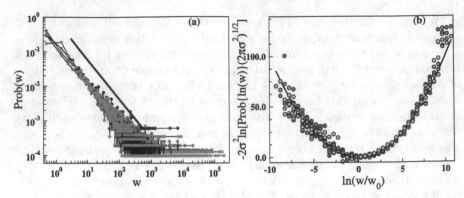

Fig. 5. Trial of power law and log-normal fits: (**a**) A double logarithmic plot of the probability distribution Prob(w) of the link weights for the six different years at the ten years interval from 1950 to 2000. The straight line shows average slope of the intermediate regime of all distributions giving an average estimate for the exponent $\tau_w = 1.22 \pm 0.15$. (**b**) Scaled plot of the probability distribution of the link weights $-2\sigma^2 \ln[\text{Prob}\{\ln(w)\}\sqrt{2\pi\sigma^2}]$ as a function of $\ln(w/w_0)$. The five year averaged data have been plotted for ten different periods from 1951 to 2000. Points scatter around the scaled form of the log-normal distribution $y = x^2$ evenly except at the ends. Contact authors for color figures.

to 2000 at the interval of ten years have been plotted with different colored symbols. Each plot has considerable noise which is more prominent at the tail of the distribution. Yet one can identify an intermediate region spanning little more than two decades of w_{ij} where the individual plots look rather straight. This indicates the existence of a power law dependence of the distribution: Prob(w) $\sim w^{-\tau_w}$ in the intermediate regime. Therefore we measured the slopes of these plots in the intermediate region for every annual ITN for 53 years from 1948 to 2000. These values have fluctuations around their means and our final estimate for the exponent is: $\tau_w = 1.22 \pm 0.15$.

We re-analyzed the same data by trying to fit a log-normal distribution as:

$$\text{Prob}(w) = \frac{1}{\sqrt{2\pi\sigma^2}} \frac{1}{w} \exp\left(-\frac{\ln^2(w/w_0)}{2\sigma^2}\right), \qquad (2)$$

where the characteristic constants constants of the distribution are defined as $w_0 = \exp(\langle \ln(w) \rangle)$ and $\sigma = \{\langle (\ln(w))^2 \rangle - \langle \ln(w) \rangle^2\}^{1/2}$. It is found that different annual ITNs have different values for w_0 and σ. However we observed that one can make a plot independent of these constants. Given the w_{ij} values of an ITN one calculates first w_0 and σ. Then calculating the Prob$\{ln(w)\}$ one plots $-2\sigma^2 \ln[\text{Prob}\{\ln(w)\}\sqrt{2\pi\sigma^2}]$ as a function of $\ln(w/w_0)$ which should be consistent with a simple parabola $y = x^2$ for all years (Note that Prob$\{\ln(w)\}d\{\ln(w)\} = \text{Prob}(w)dw$ implies Prob$\{\ln(w)\} = w\text{Prob}(w)$). This analysis has been done for fifty years for the period 1951–2000 but the data for every successive five years period have been averaged to reduce noise and ten plots for the intervals 1951–55, 1956–60, ..., 1996–2000 have been plotted in Fig. 5(b) with different colored symbols. We observe that the data points are evenly distributed around the $y = x^2$ parabola in most of the intermediate region with slight deviations at the two extremes, i.e., at the lowest and highest values of $\ln(w/w_0)$. We conclude that the probability distribution of link weights of the annual ITNs is well approximated by the log-normal distribution and is a better candidate to represent the actual functional form of the Prob(w) than a power law. We mention here that the trade imbalances have also been claimed to follow the log-normal distribution [8].

In the world wide trade relations who is stronger and who is weaker? A measure of the capacity of trade is defined by the strength s_i of a node which is the total sum of the weights w_{ij} of the links meeting at a node i. Thus,

$$s_i = \Sigma_j w_{ij}. \tag{3}$$

Using the strength distribution one can estimate which countries actually control a major share of the international trade market. We define $f_w(s)$ as the ratio of the total volume of trade a subset of countries make among themselves to the total trade volume W in the ITN. The subset is defined as those countries whose strengths are at least s. For this analysis we first arrange the nodes in a sequence of increasing strengths and then delete the nodes in this sequence one by one. When a node is deleted all links meeting at this node are also deleted. Consequently the total volume of trade among the nodes in the subset also decreases. In the Fig. 6(a) we show how $f_w(s)$ decreases with s/s_{max} for the year 2000. Up to a large value of $s/s_{max} \approx 0.01$, $f_w(s)$ effectively remains close to unity beyond which it decreases faster. It is observed that only a few top rich countries indeed trade among themselves one half of the world's total trade volume, corresponding to $f_w(s) = 1/2$. Evidently these countries are very rich and are the few toppers in the list of strengths –

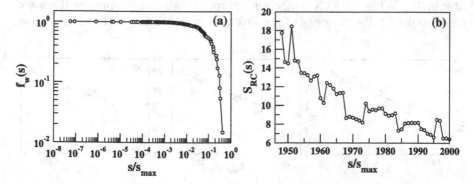

Fig. 6. Variations of the (a) fraction of the total world trade the rich-club countries make among themselves with the fractional strength of the weakest in the rich-club and (b) percentage of the world countries that make 50% of the world's total trade volume falls from $\approx 19\%$ to $\approx 6\%$ between 1948 to 2000.

which is said to have formed a 'rich-club' (RC). Therefore we measure the fraction of countries in the RC and calculate how the percentage size of the rich-club varied with time. In Fig. 6(b) we plot the year-wise fractional size S_{RC} of the rich club from 1948 to 2000 and see that it has been decreased more or less systematically from $\approx 19\%$ to $\approx 6\%$. This implies that though the world economy is progressing fast and more and more countries are taking part in the world trade market yet a major share of the total trade is being done only among a few countries within themselves.

A country makes different volumes of trade with other countries. Therefore the values of the weights associated with the links of a node, both for imports and exports vary quite a lot. A numerical measure of this fluctuation is given by the 'disparity' measure Y. For a node i the disparity is measured by

$$Y_i = \sum_{j=1}^{k_i} \left[\frac{w_{ij}}{s_i} \right]^2 . \tag{4}$$

The average disparity measure $Y(k)$ over all nodes of degree k is calculated. If the weights associated with the k links are of the same order then $Y(k) \sim 1/k$ for large k values where as if the weights of a few links strongly dominate over the others then $Y(k)$ is of the order of unity. We have measured three disparity measures, namely $Y(k)$ for the link weights w_{ij} as the total trade, $Y(k_{exp})$ for the link weights w_{ij}^{exp} as the export from the node i to node j and $Y(k_{imp})$ for the link weights w_{ij}^{imp} as the import from the node j to node i. These quantities are plotted in Fig. 7 on log-log scales and we observe power law dependences as: $kY(k) \sim k^{0.56}$, $k_{imp}Y(k_{imp}) \sim k_{imp}^{0.58}$ and $k_{exp}Y(k_{exp}) \sim k_{exp}^{0.54}$. Similar variations are also observed in trade imbalances [8].

To summarize, in this paper we have presented the analysis of the international countrywise trade data and studied the variations of different quantities associated with the International Trade Network, the trade data being available in [15]. While the ITN is inherently directed, where two opposite flows are associated with the majority of the links, we largely ignored the directedness

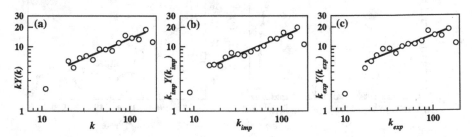

Fig. 7. Binned plot of the nodal disparity measures with degree. Averaged ITN data for the period 1991 to 2000 have been used. The best fit straight lines are (a) $kY(k) \sim k^{0.56}$ for the trade network, (b) $k_{imp}Y(k_{imp}) \sim k_{imp}^{0.58}$ for the import network and (c) $k_{exp}Y(k_{exp}) \sim k_{exp}^{0.54}$ for the export network.

and analyzed the network as an undirected graph. Our analysis shows that the link weight probability distribution of the undirected ITN fits better to a log-normal distribution as observed in [4,6]. We also show that the deviation in the size of the giant component of the ITN from the fully connected graph decays exponentially. The size of the rich-club whose internal trading amount is half of the total world trade amount decreases as time passes. Finally in the disparity measure of the ITN we distingushed between export and import at a node. It is observed that the three different disparity measures using link weights as the total trades, exports and imports grow in a similar fashion.

We thank A. Chatterjee and B.K. Chakrabarti for their nice hospitality in the ECONOPHYS-KOLKATA III meeting.

References

1. Albert R and Barabási A L (2002) Statistical Mechanics of Complex Networks. Rev. Mod. Phys. 74:47
2. Dorogovtsev S N and Mendes J F F (2003) Evolution of Networks, Oxford University Press
3. Newman M E J (2003), SIAM Review 45:167
4. Serrano M Á and Boguñá M (2003) Phys. Rev. E. 68:015101
5. Garlaschelli D and Loffredo M I (2004) Phys. Rev. Lett. 93:188701
6. Serrano M Á (2006) physics/0611159
7. Bhattacharya K, Mukherjee G, Saramäki J, Kaski K and Manna S S (2007) The International Trade Network: weighted network analysis and modelling, preprint
8. Serrano M Á and Boguñá M and Vespignani A (2007) arXiv:0704.1225
9. Deo N (1995) Graph Theory with Applications to Engineering and Computer Science, Prentice-Hall of India
10. Barrat A, Barthélemy M, Pastor-Satorras R, and Vespignani A (2004) Proc. Natl. Acad. Sci. USA 101:3747
11. Barrat A, Barthélemy M and Vespignani A (2005) J. Stat. Mech. P05003
12. Mukherjee G and Manna S S (2006) Phys. Rev. E 74:036111
13. Onnela J P, Saramäki J, Hyvönen J, Szabó G, Lazer D, Kaski K, Kertész J, and Barabási A L (2006) physics/0610104
14. Human Development Reports in http://hdr.undp.org/
15. Version 4.1 in Expanded Trade and GDP Data // Kristian Skrede Gleditsch http://privatewww.essex.ac.uk/~ksg/exptradegdp.html
16. Tinbergen J (1962) An Analysis of World Trade Flows in Shaping the World Economy, Tinbergen, ed. New York: Twentieth Century Fund
17. Gleditsh K S (2005) private communication

Networks of Firms and the Ridge in the Production Space

Wataru Souma

NiCT/ATR CIS Applied Network Science Lab., Kyoto, 619-0288, Japan
souma@nict.go.jp, souma@atr.jp

We develop complex networks that represent activities in the economy. The network in this study is constructed from firms and the relationships between firms, i.e., shareholding, interlocking directors, transactions, and joint applications for patents. Thus, the network is regarded as a multigraph, and it is also regarded as a weighted network. By calculating various network indices, we clarify the characteristics of the network. We also consider the dynamics of firms in the production space that are characterized by capital stock, employment, and profit. Each firm moves within this space to maximize their profit by using controlling of capital stock and employment. We show that the dynamics of rational firms can be described using a ridge equation. We analytically solve this equation by assuming the extensive Cobb-Douglas production function, and thereby obtain a solution. By comparing the distribution of firms and this solution, we find that almost all of the 1,100 firms listed on the first section of the Tokyo stock exchange and belonging to the manufacturing sector are managed efficiently.

1 Introduction

Network science has been used to clarify many characteristics of networks in the real world, from biological networks to the worldwide web. This is also used in the field of econophysics. Such studies have investigated business networks [9], shareholding networks [6, 10–12], world trade networks [7, 8], corporate board networks [3, 5], and transaction networks [13]. However, we have little knowledge about networks in the economy. This is because previous studies have focused on one type of relationship between nodes. In the real world, for example, firms are connected to each other by a wide variety of relationships such as shareholding, interlocking directors, transactions, etc. Networks and graphs that incorporate many kinds of connections, i.e., multi-edges, are called multigraphs. Thus, the purpose of this article is to look at networks in the economy as multigraphs.

The economy is regarded as a set of activities of irrational agents in complex networks. However, many traditional studies in economics have investigated the activities of rational and representative agents in simple networks, i.e., regular networks and random networks. To overcome the limitations of such an unrealistic situation, the viewpoint of irrational agents has emerged. However, it is important to reconsider how rational firms behave in a production space. Here, the production space is composed of capital K, labor L, and profit, Π, and Π is a function of K and L. In this production space, rational firms are managed to maximize profit by control of the capital and the labor [1]. Thus, the second purpose of this article is to find a law followed by rational firms in the production space.

This paper is organized as follows. In the next section we define networks in the economy as multigraphs. Here, the nodes of the multigraph are firms, and the relationships between those firms are shareholding, interlocking directors, transactions, and joint applications for patents. For this multigraph, we calculated a variety network indices, and clarified the characteristics of said networks. In the following section, we discuss the efficient management of firms. We propose that the efficient firms must climb the mountain of the profit by climing along a ridge of that mountain. The last section is devoted to a summary.

2 Multigraph in the Economy

When we consider networks in the economy, we must define nodes and edges. In the case of the economy, nodes are things like individuals, groups, firms, and countries. We use Japanese firms listed on the stock market in 2004 as nodes. The total number of firms is 3,576. Though there are many relationships between these firms, in this article, we consider only four types of relationships: shareholding, interlocking directors, transactions, and joint applications for patents. These edges are regarded as multiedges, and networks are regarded as multigraphs. We evaluate the number and type of edges and use this to assign a weight to those edges. For example, if two firms are connected by one instance of all of these four relationships, we consider these firms are connected by one edge with a weight of 4. Thus, the multigraph is a weighted graph.

2.1 Multigraphs in the Automobile Sector

As an example of a multigraph, we have developed a network made up of 13 firms in the automobile sector. This network is shown in Fig. 1. In this figure, a number beside an edge represents the weight of that edge.

We calculated a variety of indices that characterize the network. Indices calculated for each firm are summarized in Table. 1. In this table, k_i is the

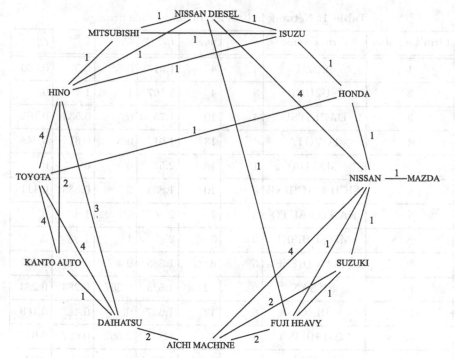

Fig. 1. Multigraph of 13 firms in the automobile sector

degree of node i, and w_i is the sum of the weight of the edges connected to the node i:

$$w_i = \sum_{j \in a(i)} w_{ij},$$

where $a(i)$ represents a set of nodes connecting to node i, and w_{ij} is the weight of the edge that joins node i and node j. Here, i denotes a firm's number as shown in Table. 1. If we look at k_i and w_i, we can see that NISSAN and HINO have the greatest value of the degree, and that TOYOTA has the greatest value of the sum of the weight of the edges. From this result, we can see that firms with the greatest value of the degree do not necessarily have the greatest value of the sum of the weight of the edges.

In Table. 1, L_i is the averaged path length of node i. The averaged value of L_i is $\langle L \rangle = 1.75$. From Fig. 1, so we can expect MAZDA to have the greatest value of the averaged path length. In this case, this expectation is correct, and MAZDA has $L_5 = 2.5$. However, this value is not much greater than the values of SUZUKI and KANTO AUTO, which are both $L_2 = L_7 = 2.167$. As we can see in Fig. 1, MAZDA is connected to other nodes in the network through NISSAN, and NISSAN has the lowest value of the averaged path length at $L_{10} = 1.583$. This is why MAZDA does not have a very high value of the averaged path length.

Table 1. Network indices for 13 automobile firms

Firm's number	Firm's name	k_i	w_i	L_i	C_i	C_i^{w}	C_i^{b}
1	ISUZU	4	4	1.917	0.5	0.5	0.030
2	SUZUKI	3	4	2.167	1	1	0
3	DAIHATSU	4	10	1.75	0.5	0.533	0.105
4	TOYOTA	4	13	1.917	0.5	0.615	0.048
5	MAZDA	1	1	2.5	0	0	0
6	AICH MACHINE	4	10	1.833	0.5	0.533	0.111
7	KANTO AUTO	3	7	2.167	1	1	0
8	MITSUBISHI	3	3	2	1	1	0
9	NISSAN DIESEL	5	8	1.583	0.4	0.344	0.197
10	NISSAN	6	12	1.583	0.267	0.283	0.283
11	HINO	6	12	1.667	0.4	0.4	0.160
12	FUJI HEAVY	4	5	1.833	0.667	0.667	0.042
13	HONDA	3	3	1.75	0	0	0.085

In this table, C_i is the clustering coefficient of node i, and C_i^{w} is the weighted clustering coefficient of it. These are defined as

$$C_i = \frac{2e_i}{k_i\,(k_i - 1)}, \tag{1}$$

and

$$C_i^{\mathrm{w}} = \frac{1}{w_i\,(k_i - 1)} \sum_{j,k} \frac{w_{ij} + w_{ik}}{2}, \quad \text{if } j \text{ and } k \text{ are connected,}$$

respectively. Here, e_i is the number of edges which connect the nodes neighboring of node i. As we can see in this table, if we ignore the weights of the edges, ISUZU, DAIHATSU, TOYOTA, and AICH MACHINE have the same value of clustering coefficient, which is $C_1 = C_3 = C_4 = C_6 = 0.5$. However, if we consider the weight of the edges, we can see a difference. In this case, TOYOTA has $C_4^{\mathrm{w}} = 0.615$, which is greater than C_1^{w}, C_3^{w}, and C_5^{w}.

In Table 1, C_i^{b} is a betweenness centrality of node i, and is defined as

$$C_i^{\mathrm{b}} = A \sum_{i \neq j \neq k \in V} \frac{\sigma_{jk}(i)}{\sigma_{jk}},$$

where A is a normalization constant, and V represents a set of nodes in the network. Here, σ_{jk} is the number of the shortest paths from node j to node

k, and $\sigma_{jk}(i)$ is the number of the shortest paths from node j to node k through node i. In this table, we can see that NISSAN has the greatest value of betweenness centrality, at $C_{10}^{\mathrm{b}} = 0.238$. As mentioned previously, MAZDA is connected to the network through NISSAN. Thus, firms connected to NISSAN are thereby also connected to MAZDA. This is why NISSAN has the greatest value of betweenness centrality.

2.2 Network Indices for Various Sectors

We calculated network indices for the largest connected component (LCC) in networks of a variety of industry sectors. The results are summarized in Table. 2. In this table, $\langle k \rangle$, $\langle L \rangle$, $\langle C \rangle$, and $\langle C^{\mathrm{w}} \rangle$ are the averaged values of the degree, the averaged path length, the clustering coefficient, and the weighted clustering coefficient.

By considering all industry sectors, we obtained the results shown in the second column in the table. In this case, the size of the largest connected component is the same as the network itself. This means that there is no disconnected component. By considering the electronics sector, we obtained the result shown in the third column in the table. By comparing the second and third columns, we can see remarkable differences between the values of $\langle k \rangle$, $\langle C \rangle$, and $\langle C^{\mathrm{w}} \rangle$. The network of all sectors has a greater value of $\langle k \rangle$ than that of the electronics sector. Thus, we expect the network of all sectors to have a greater value of $\langle C \rangle$ and $\langle C^{\mathrm{w}} \rangle$ than that of the electronics sector. However, this is not true in this case. From Eq. (1), we can see that the average node has a clustering coefficient given by

$$C_i = \frac{2e_i}{\langle k \rangle (\langle k \rangle - 1)}.$$

Table 2. Network indices for some business sectors

Indices	All sectors	Electronic	Automobile	Medical	Steel
#node	3,576	285	87	49	47
#link	55,332	2,196	569	126	119
#node(LCC)	3,576	285	85	49	47
#link(LCC)	55,332	2,196	568	126	119
$\langle k \rangle$	29.723	15.411	13.365	5.143	5.064
$\langle L \rangle$	2.777	2.319	1.984	2.528	2.300
$\langle C \rangle$	0.240	0.652	0.890	0.220	0.571
$\langle C^{\mathrm{w}} \rangle$	0.187	0.452	0.461	0.164	0.417

Thus, if different nodes have almost the same value of e_i, the great value of $\langle k \rangle$ makes C_i small. This is why the network of all sectors has a lesser value, C_i, than that of the electronics sector. This is also applicable to the case of $\langle C^{\mathrm{w}} \rangle$.

By considering the automobile sector, we obtained the results shown in the fourth column in the table. As we can see, the electronics and automobile sectors have almost the same value of $\langle k \rangle$ and $\langle L \rangle$. However, the value of $\langle C \rangle$ in the automobile sector is markedly greater than that in the electroniics sector. This may lead us to believe that the automobile sector has stronger connections than the electronics sector. However this is not the case, because the value of $\langle C^{\mathrm{w}} \rangle$ is almost the same in both the electronics and automobile sectors.

The sizes of the electronics and the automobile sectors' networks differs. Thus, it is important to compare sectors of about the same size. The medical and steel sectors are such a case. As we can see in the fifth and sixth columns in the table, these two sectors have almost the same number of nodes and edges. Thus, the $\langle k \rangle$s are almost the same value. Although $\langle L \rangle$s are almost same, $\langle k \rangle$s and $\langle L \rangle$s are very different. Both the $\langle C \rangle$ and $\langle C^{\mathrm{w}} \rangle$ in the steel sector are greater than those in the medical sector. This means that the steel sector is constructed from strongly connected clusters, compared to the weaker connected clusters in the medical sector.

The correlation between the degree and the betweenness centrality is shown in Fig. 2. This figure is the log-log plot of the degree, k, and the

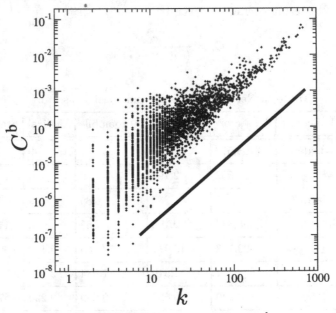

Fig. 2. Correlation between k and C^{b}

betweenness centrality, C^b. In this figure, the bold solid line corresponds to $C^b \propto k^2$. As we can see, this relation is approximately correct.

3 A Ridge in the Production Space

Production is one of the characteristics of a firm's activity. The original concept of production is simple: two inputs, capital, K, and labor, L, are combined to produce a unique maximum quantity of one output, known as production, Y (for example see Ref. [14]). This is represented by

$$Y = F(L, K), \ L > 0, \ K > 0.$$

The function, F, is called a production function and defines the technical relationship between the two inputs and the output. If the production function satisfies

$$Y = Lf(x), \quad x \equiv \frac{K}{L},$$

then the production function is called "extensive". Profit, Π, of firms is defined by

$$\Pi = Y - D - L.$$

Here, we assume that Y is replaced by the extensive production function $Lf(x)$. In addition, we assume that D is represented by rK, where r is a debt interest rate. Therefore we obtain

$$\Pi = L\{f(x) - rx - 1\} \equiv L\pi(x) \tag{2}$$

We also assume that the extensive production function is of a Cobb-Douglas type [4], i.e.,

$$f(x) = Ax^\alpha, \tag{3}$$

where $A = 1.2278$ and $\alpha = 0.1365$ for 1,100 firms, all of which are listed on the first section of the Tokyo stock exchange (TSE), and belong to the manufacturing sector. The averaged value of the debt interest rate of these firms is $\bar{r} = 0.00187$. We use this value below.

Contours in the profit space are shown in Fig. 3. In this figure, the horizontal axis corresponds to L and the vertical axis corresponds to K. The thin solid lines represent the contours in the profit space, and each dot corresponds to each firm. Each firm climbs the mountain of profit to maximize their profit. It is reasonable to consider that an efficient firm adopts the direction of a gradient vector:

$$\nabla \Pi = \left(\frac{\partial \Pi}{\partial L}, \frac{\partial \Pi}{\partial K} \right).$$

Thus, a trace of the movement of the most efficient firm makes the steepest-ascent line.

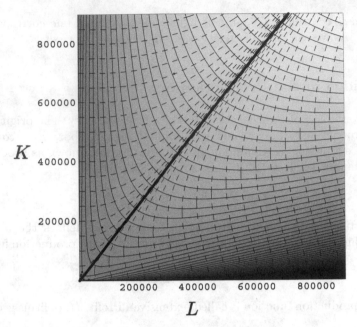

Fig. 3. Contour plot of the profit space

In Fig. 3, the dashed lines correspond to the steepest-ascent lines. As we can see, these steepest-ascent lines converge in a confluent line. We call this confluent line a ridge. The ridge maximizes the length of the gradient vector on the same contour. This is obtained as an extreme solution of

$$\frac{1}{2}\left(\boldsymbol{\nabla}\Pi\right)^2 - \lambda\left(\Pi - \Pi_0\right),$$

where λ is Lagrange's unknown constant and Π_0 is another constant. This equation is rewritten as "the ridge equation" [2]:

$$\sum_{j=1,2}\frac{\partial^2\Pi}{\partial q_i\partial q_j}\frac{\partial\Pi}{\partial q_j} = \lambda\frac{\partial\Pi}{\partial q_i}, \quad (i=1,2), \tag{4}$$

where $(q_1, q_2) \equiv (L, K)$.

If we insert Eq. (2) and Eq. (3) into Eq. (4), we can obtain an eigenvalue equation:

$$\begin{pmatrix} x^2 & -x \\ -x & 1 \end{pmatrix}\begin{pmatrix} \pi - x\pi' \\ \pi' \end{pmatrix} = \tilde{\lambda}\begin{pmatrix} \pi - x\pi' \\ \pi' \end{pmatrix},$$

where a prime represents a derivative with respect to x. Here, we define $\tilde{\lambda} \equiv L\lambda/\pi''$.

There are two solutions. One is $x\pi - (x^2 + 1)\pi' = 0$ for $\tilde{\lambda} = 0$, and the other is $\pi = 0$ for $\tilde{\lambda} = x^2 + 1$. However the latter solution, i.e., $\pi = 0$, is

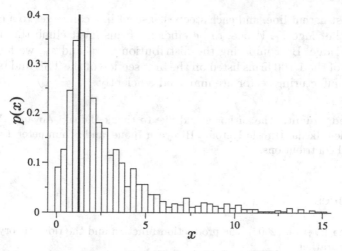

Fig. 4. Distribution of $x \equiv K/L$

trivial, because this means no profit. Thus, we are interested in the former solution. If we solve the former equation, we can obtain $x = 1.2597 \equiv x_r$. This is represented by the bold and solid lines in Fig. 3.

To clarify the difference between x_r and x_i, we consider a probability distribution of x_i. This distribution is shown in Fig. 4. In this figure, the bold solid line corresponds to x_r. In this figure, we can see that the peak of the distribution and x_r almost coincide. This means that almost all of the 1,100 firms listed on the first section of the TSE and belonging to the manufacturing sector are managed efficiently.

4 Summary

The first part of this paper looks at networks in the economy, and the second part of this paper is about the efficient behavior of firms in the production space. In the first part of this paper, we considered networks of the economy as a multigraph, and treated the multigraph as a weighted network. We clarified some of the characteristics of the networks in some industry sectors by calculating various network indices. There are two interesting observations: the clustering pattern depends strongly on the sector, and the degree and the betweenness centrality correlate strongly and have approximately the relation, $C^b \propto k^2$.

In the second part of this paper, we considered the dynamics of firms in the production space. These dynamics are evaluated on the basis of a firm's capital, K, labor, L, and profit, Π, and Π is a function of K and L. This space is called the mountain of the profit. Firms climb this mountain to maximize their profit by control of K and L. Thus, the trace of movement of firms makes the steepest-ascent line, and each steepest-ascent line converges to a confluent line, i.e., the ridge x_r. Hence, to be efficient, firms must climb the mountain along the ridge. By comparing the distribution of x_i and x_r, we found that almost all of the 1,100 firms listed on the first section of the TSE and belonging to the manufacturing sector are managed efficiently.

Acknowledgement. The author would like to thank Hideaki Aoyama, Yoshi Fujiwara, Yuichi Ikeda, Hiroshi Iyetomi, Hiroyasu Inoue, and Schumpeter Tamada for their useful contributions.

References

1. Aoyama H, et al. (2007) The production function and the ridge theory of firms. in preparation
2. Aoyama H, Kikuchi H (1992) A new valley method for instanton deformation. Nuclear Physics B 369: 219–234
3. Battiston S, Bonabeau E, Weisbuch G (2003) Decision making dynamics in corporate boards. Physica A 322: 567–582
4. Cobb CW, Douglas PH (1928) A theory of production. American Economic Review 18: 139–165
5. Davis G, Yoo M, Baker WE (2003) The small world of the American corporate elite, 1982–2001. Strategic Organization 3: 301–326
6. Garlaschelli D, et al. (2003) The scale-free topology of market investments. Physica A 350: 491–499
7. Li X, Jin YY, Chen G (2003) Complexity and synchronization of the World trade Web. Physica A 328: 287–296
8. Li X, Jin YY, Chen G (2004) On the topology of the world exchange arrangements web. Physica A 343: 573–582
9. Souma W, Fujiwara Y, Aoyama H (2003) Complex networks and economics. Physica A 324: 396–401
10. Souma W, Fujiwara Y, Aoyama H (2004) Random matrix approach to shareholding networks. Physica A 344: 73–76
11. Souma W, Fujiwara Y, Aoyama H (2005) Heterogeneous economic networks. In: Namatame A, et al. (Eds.) The Complex Networks of Economic Interactions: Essays in Agent-Based Economics and Econophysics (Lecture Notes in Economics and Mathematical Systems, Vol. 567), Springer-Verlag, Tokyo, pp. 79–92
12. Souma W, Fujiwara Y, Aoyama H (2005) Shareholding networks in Japan. In: Mendes JFF, et al. (Eds.) Science of Complex Networks: From Biology to the Internet and WWW CNET2004 (API Conference Proceedings, Vol. 776), Melville, New York, pp. 298–307.
13. Souma W, et al. (2006) Correlation in business networks. Physica A 370: 151–155
14. Varian H.R (1992) Microeconomic Analysis – Third edition. W. W. Norton & Company, Inc.

Debt-credit Economic Networks
of Banks and Firms: the Italian Case

Giulia De Masi and Mauro Gallegati

Dipartimento di Economia, Università Politecnica delle Marche, P.le Martelli 8, 60121 Ancona, Italy g.demasi@univpm.it,m.gallegati@univpm.it

Summary. A first analysis of the Italian system of banks and firms was carried out using an approach based on network theory. The emerging architecture of this economic network shows peculiar behaviors. Big banks are creditors of many firms; among these, big firms are financed by several banks. On the contrary, small firms are preferentially financed by small banks, covering very often the entire credit they need.

1 Introduction

A complete understanding of the architecture of credit relationships in economic systems is of primary importance. The problem of the economic stability in fact is strongly related to the underlying structure of credit/debt relationships among its components: this structure plays a crucial role in the exposure to risk of avalanche failures and domino effects in systems characterized by credit relationships [1]. In general three kinds of credit relationships can be identified: commercial credit among firms, financial credit from banks to firms and inter-bank credit. In the following we focus on the second type of credit relationships. The presence of credit/debt relationship allows us to define a network of credit where nodes are the economic agents (banks and firms) and the links are credit relationships.

The firms behave very differently from each other. Some firms obtain credit just from one bank, others have loans from many banks (multiple relationships). Given the relevance of the problem, a recent literature focuses on this topic [2–4].

In literature the main questions are: What determines the number of bank relationships? What kind of firms prefer multiple links? Why do they prefer multiple relations? Is there an "optimal" number of bank relationships?

In this paper we provide a first analysis of the Italian data of credit relationships between banks and firms using networks tools. This paper differs from previous studies not merely for the method used, but because the network analysis allows to obtain information on the structure of relationships and in particular in the architecture of second order or higher. While standard econometric analysis are useful to detect correlations among different features of banks and firms, they do not allow to study the topology of the underlying architecture of bank-credit relationships. Nevertheless the structure of relationships plays a crucial role in bankruptcy diffusion. We organize the paper in the following way

- Sect. 1: introduction
- Sect. 2: an overview of the problem of multiple relationships in literature
- Sect. 2.1: the specific case of Italy
- Sect. 3: description of dataset
- Sect. 4: network approach
- Sect. 5: networks of Italian banks and firms: statistical characterization
- Sect. 6: network of cofinancing banks
- Sect. 7: discussion and conclusions

2 Multiple vs Single Bank Relationships

In recent years, the exploration of the structure of credit relationships among banks and firms has acquired increasing importance [5]. In fact the possibility to obtain big storage of data allows to study the credit relationships of a large set of firms with banks in different countries [3].

The study of single-multiple relationships has been carried out for many countries, as Germany [6], Japan [2], Portugal [7]. The number of bank-relationships strongly depends on the kind and the size of sample under study. A comparison among different countries shows that Italy represents an extreme case with an average of 15.5 banks per firm, maintaining four more bank relationships per firm than the country with the second highest number of relationships, Portugal (11.5 banks per firm). Only 2.9% of Italian firms has single bank relationship; At least 95% of firms in Portugal, Spain, France, Belgium, Greece, the Czech Republic and Finland maintain a relationship with more than one bank: on average they have 8 banks per firm. In UK the average number of banks is 2.6 and 25% of firms maintain only one bank relationship [3].

In some countries, such as in Japan and Germany, among all the banks, a firm is particularly influenced by a certain bank (the inside bank), when the bank maintains a share of the capital of that firm. In these cases the inside bank has a better access to information (monopolistic information [2]) on the actual financial condition of the firm.

The main aspects playing a role on the multiple credit relationships has been investigated in a very recent paper by Ogawa et al. [2]. The theory of the optimal number of banks relationships is a very complex field of research given the high number of advantages and disadvantages in the choice both of single relationship and multiple relationships. On the firm side, the existence of a single bank relationship is based on the minimization of costs of transactions and monitoring; on the contrary, firms can try benefit in a competing market of banks: this implies a growth of number of relationships. Multiple bank configuration guarantees the firm against liquidation risk: among many banks, at least one will be able to face a premature liquidation. From the bank side, a bank prefers to finance firms having multiple bank relationships in order to pool the risk of failure of the firm. On the other hand, single linkage gives a bigger control to the bank on the financial choices of the firm.

From a theoretical point of view, in recent years the development of information theory has renewed the interest to borrowing relationships, as an example of mechanism of informational asymmetry between creditor and debtor. The amount of information plays a crucial role in the competition among banks to provide loans to firms. Among banks, there are two kinds of competitions: the first one is on the market of credit and is driven by the interest rate, the second one is among banks financing firms [8].

There is some evidences, in particular related to the system of Portuguese firms-banks, that some firms start with a single relationship and after some time it switch to multiple links in conditions of growth opportunities, more banks debt and less liquidity [7].

2.1 The Italian Case

We have already observed that, among European countries, Italy has the biggest average number of banks relationships per firm. The multiple bank relationships became more relevant in Italy in the 50s [9]. In fact after the second world war Italian firms were able to finance themeselves: therefore the interaction among banks and firms was very small. In the 70s the worsening of financial conditions of Italian firms induced entrepreneurs to ask for credit to the banks. In these years the multiple bank relationships emerged. On the one hand this implied that a firm received credit from more than one bank, on the other hand that implied that a good knowledge of the real economic conditions of firm was not easily available to banks. In many cases credit was provided also for personal trust reasons, without a deep investigation on the actual financial condition of the firm [9].

Detragiache et al. studied data of medium-small size Italian manufacturing firms (firms with less than 500 employees). In 1994, 89% of Italian firms have multiple links; the median number of relationship is 5 and the 75th percentile is 8 [4]. The phenomenon of multiple linking is more striking also respect to U.S.. Pagano et al. studied a sample of the biggest 2181 Italian companies from 1982 to 1992 (companies with shareholder's equity bigger than 5 billion lire) and refer that the mean number of relationships is 13.9 and the median of the number of banks is 12 [10].

While it is quite clear why big firms have multiple links, it is not clear why small firm have multiple links [8].

3 Data Set

In this work we analyze a subset of the AIDA database [11]. We study data regarding Italy. This set contains information on the largest 170,000 Italian societies from 1992. We have detailed information on the characteristics of each firm, as the total net worth, the total asset, the solvency ratio, the number of workers, the added value. Moreover we know the identity of banks financing each firm. This allows us to study how the number of relationships varies for firms of different size. We merge this dataset with another one derived from Bank of Italy classification of the whole set of Italian banks in 5 groups [12]: large banks, medium banks, small banks, cooperative credit banks and rural banks. In this way we can identify the attributes of the banks lending to each individual firm and to detect different behavior of banks of different sizes. In the analysis of this second database we had to take in consideration phenomenon of mergers and acquisition of banks in the years of the sample (this is a relevant phenomenon in Italy, due to the necessity of competitiveness in international markets. For instance the number of banks passed from 1073 to 814 from 1992 to 2002). The Bank of Italy provides a classification of Italian banks in 5 categories:

- larger banks have funds of more than 45 billion Euros
- large banks between 20 and 45
- medium between 7 and 20
- small between 1 and 7
- the remaining banks are minor banks.

Our sample is composed by 11 larger banks, 11 large banks, 34 medium banks, 125 small banks and 307 smaller banks. In this paper we focus on credit relationships of Italian firms with Italian banks in the year 2004. The set is composed by 55005 contracts among 488 Italian banks and 33468 Italian firms. The study of the evolution of the structure of the credit network will appear in another paper [13].

We distinguish firms following the ISTAT classification [14]: in 2004 the Italian firms were 4.1 million: the 95% micro, 4.5% small, 0.5% medium. In our sample 29% are micro, 40% small, 23% medium, 7% large [15]. In this sense our sample privileges the analysis of big firms and does not consider small firms and small banks.

Unfortunately we do not have access to the amount, the rate and the duration of credit, but this dataset is more extended than others studied in the literature.

4 The Network Representation

We represent this system as a network, in order to use an approach based on the graph theory to analyze the complex structure of credit relationships in the Italian economic system.

A network is defined as a set of nodes and links. It is mathematically represented as a graph. In recent years a big development of the complex networks theory has been observed. Many real systems have been represented as networks. This kind of approach allows to get insights into the architecture of complex systems composed by many objects, linked to each other in a non trivial way [16,17]. Most of them show peculiar scaling properties. In particular many networks are scale-free, that means that the degree distribution is power-law tailed. Applications of network theory in economics can be useful in order to consider explicitly the relations among economic agents [18]. Many empirical analysis of economic systems have be done using networks tools: main examples are the applications to world trade web [19, 20], interbank market [21], stock market [22], e-commerce [23].

In our particular case the nodes are banks and firms [27]. The links represent the credit relationships among them. These kinds of networks, composed by two kinds of nodes, are called "bipartite networks".

Moreover we can extract two networks from the overall network, each one composed by just one kind of node: this method is called one-mode reduction and the two networks "projected networks", in the sense that they are obtained as a projection of the initial graph in the subspace composed by only one kind of node.

A network is represented from a mathematical point of view by an adjacency matrix. The element of the adjacency matrix a_{ij} indicates that a a link exists between nodes i and j, that is $a_{ij} = 1$ if the bank i provides a loan to the firm j; otherwise $a_{ij} = 0$.

The degree of a node i is the number of links outgoing from it and is calculated by

$$k_i = \sum_j a_{i,j} \qquad (1)$$

The assortativity is a measure of similarity among nodes and it is defined as

$$k_{nn}(i) = \frac{1}{k_i} \sum_{j \in \mathcal{V}(i)} k_j. \qquad (2)$$

The clustering coefficient c_i is a measure of the density of connections around a vertex and is defined as

$$c_i = \frac{2}{k_i(k_i - 1)} \sum_{j,h} a_{ij} a_{ih} a_{jh}. \qquad (3)$$

Hence, the clustering coefficient allows to calculate the proportions of the nearest neighbors of a node that are linked to each other. The average clustering coefficient,

$$C = \frac{1}{N} \sum_i c_i$$

expresses the statistical level of cohesiveness measuring the global density of interconnected vertex triplets in the network.

In a graph, the distance between two vertices is defined as the length of the shortest path joining them. In a connected graph the average distance is the average over all distances. If the graph is not connected, the average distance is defined as the average among all distances for pairs both belonging to the same connected component. The diameter of a graph is given by the maximum of all distances between pairs.

Many tentative of studies have been done in the field of bipartite graphs [24–26]. While we can apply the above measures in networks where all nodes are of the same kind, in bipartite networks some of these measures can not be applied, because of the different nature of the nodes. Therefore the main statistical quantities under study are degree distribution of each of the two kinds of nodes and correlations among the degree of the two kinds of nodes. A measure of connectedness in bipartite networks is given by the density of cycles of size 4 surrounding a node [28].

In the case of projected networks, the quantities under study are degree distribution, clustering coefficient, assortativity, betweenness centrality, diameter [24].

In the following we apply these tools to our network.

5 Total Network

We denote firms with the index f and banks with the index b.

By definition each bank is linked just with firms and each firm is linked just with banks.

To represent the network, we should decide on a criterium because the whole network is too big to be represented. Therefore we select the firms with degree equal or higher than 4 and we plot all their links: the resulting network is composed by 28 firms and 43 banks with 146 links (see Fig. 1). Firms are the black dots. Banks are colored, using the Bank of Italy classification: red nodes are larger banks, orange large banks, yellow medium banks, green small banks, blue smaller banks. We can observe the single firms are connected to banks of different colors, that is belonging to different classes of bank size, ranging from larger banks to smaller banks. We can understand that this system is neither assortative nor disassortative: the most connected firms are connected both with slightly connected banks (small banks) and with highly connected banks (large banks). A more complex structure emerges, which we will study further on. There is therefore a differentiation in borrowing relationships, as emerges from the heterogeneity of size (color) of the banks providing loans to the same firm (as we will prove quantitatively below). Moreover, even if we will observe that on average small banks have less connections than big banks, there are sensible fluctuations: in particular there are small banks with an high number of contracts: presumably small banks imitate large banks and,

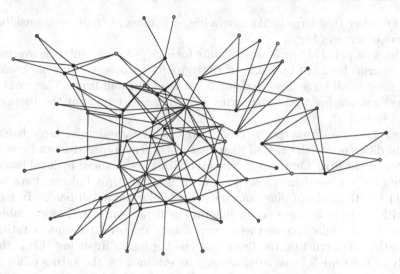

Fig. 1. Visualization of a part of the network of banks and firms (2004).

not having a wide system of rating, they finance some firms, just trusting big banks or the same firm reputation [9].

In Fig. 2 is plotted the distribution of degree for banks and firms for the year 2004.

The largest k_b is 3336, whereas the largest k_f is 8. Only one firm receives credit from 8 banks; only two receive credit from 6 banks. The behavior of banks is more heterogeneous than that of firms: in fact the degree ranges in three orders of magnitude from 1 to 3336. We observe a fat-tailed degree distribution of the degree of the banks.

In Fig. 3 left panel, we plot the average degree of banks for each group. Remembering that group 1 corresponds to larger banks and group 4 to small

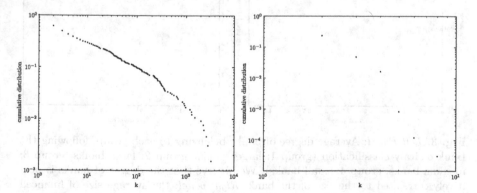

Fig. 2. Frequency distribution banks degree k_b and firms degree k_f.

banks, we notice that large banks have a bigger number of credit relationships, as we can observe in Fig. 3.

In the right panel of Fig. 3, we calculate for each group of banks the average capital of firms financed by banks of that group. We observe that big banks finance most of all large firms and small banks fund small firms. This can be explained considering that small firms find financing banks on the basis of personal trust.

In general large firms receive credit from many large banks: large banks prefer to differentiate the risk of financing just a fraction of the total credit requested by the firm. On the contrary small firms are financed by local banks and often by just one bank. Moreover, firms with multiple linkages have relationships with banks of different kinds: big banks and small banks. In fact firms with multiple linkages are in general big firms, considered more stable: therefore small banks finance them, even if they do not have done a rating investigation, just trusting the firms and the big banks financing them (the stability of the firm is considered already ascertained by the rating office of big banks). Small firms have difficulty to be financed and they are normally financed by small local banks.

Therefore, as an effect of average, the average degree of financing banks seems to not depend on the degree of the firm. This network is neither assortative nor disassortative: simple measures of scaling of the average degree of the neighbors of nodes vs their degree (commonly called degree-mixing) is not useful to understand the structure: in fact firms with a high degree have neighbors with such a heterogeneous degree that the average is not representative: in the left panel of Fig. 4 we observe a flat scaling of the average degree of financing banks versus financed firm degree k. This is just an effect

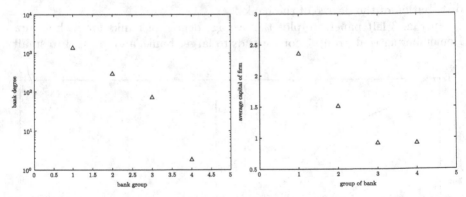

Fig. 3. *Left Panel*: Average degree of banks belonging to each group, following the Bank of Italy classification (group 1: larger banks, group 2: large banks; group 3: medium banks, group 4: small banks). We observe that the degree of banks is positively correlated to the size of the bank. *Right panel*: the average size of financed firms for each group of banks. Small banks tend to finance small firms and large firms tend to finance large banks.

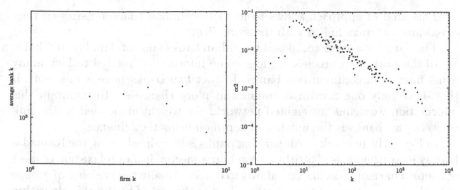

Fig. 4. *Left panel*: average degree of banks financed by firms with degree k (*left*); *right panel*: average density of loops of size 4 vs degree.

of average on the highly heterogenous degree of financing banks. In multiple relationships a firm can differentiate the kind of banks to ask for credit, obtaining loans from banks of different kinds (larger banks and minor banks, for example). The correlation between the degree of firm with the degree of its neighbors is indeed very low (0.1!).

Considering firms with degree equal to or greater than 3 (a sample of 543 firms), the average number of links with banks of groups from 1 to 5 is: 1.9; 0.8; 0.1; 0.7; 1.1. A big role is played by very small banks, financing presumably small firms. For instance, if a firm has at least one relationships with banks of group 1, it has on average a number of links with each of the other groups equal to: 0.8, 0.1, 0.06, 1.0.

In bipartite graphs, the clustering coefficient is 0: therefore we focus on the presence of loops of size 4 [28]. This measure allows to have infomation about the presence of banks financing the same firms. In the rigth panel of Fig. 4 we plot the scaling of the average density of loops of size 4 vs degree: banks with low degree (small banks) have more common firms to fund.

6 Co-financing Banks Network

In the study of bipartite graph a very widely used approach is to study separately two networks that can be defined from the original network. If we call the two kinds of nodes as nodes A and B, we can study the network G_{A+B} which has the total set of nodes $(A + B)$ or the networks G_A and G_B which have only nodes of kind A or B respectively [29, 30].

In our case we can define the network of banks and the network of firms. The first is the network of cofinancing banks: two banks are linked if they finance the same firm. The second is the network of co-financed firms, that is two firms are connected if they are financed by the same bank. The network of firms is highly disconnected; therefore we focus just on the bank network.

This kind of approach allows to identify common characteristics (if they exist) among banks linked with the same firm.

The bank network is composed by Italian banks and total undirected links.

In the projection process, we lose some information related to how many firms finance in common two banks. In fact two banks have a link both if they have only one firm in common and more than one. To maintain this information we define a weighted network: the weight associated to the link between two banks is the number of common firms they finance.

In Fig. 5 the network of cofinancing banks is visualized, using the Kamada-Kaway algorithm: this algorithm, based on a mechanism of relaxation of a set of coupled harmonic oscillators, allows to detect visually the presence of groups of banks more connected to each other than to the rest of banks: this algorithm in fact put nodes of the same group near to each other in the 2-dimensional space. We observe the presence of three communities. Each community is composed by banks of different size. Further analysis will focus on components of these communities. Disconnected nodes are banks that are reported for just one firm in our dataset. We notice that heterogeneous banks (banks of different sizes) are connected, meaning that banks of different sizes finance the same firm.

In the bank network the average degree is 5.7, while the average clustering coefficient is 0.3. The distributions of the degree and of the weights are

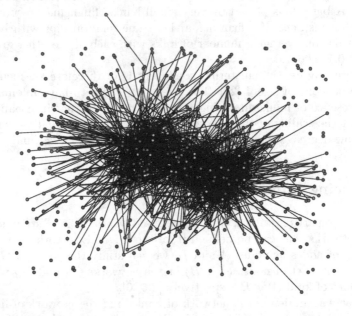

Fig. 5. Graph representation of the one-mode reduction network on the subspace of banks. The convention for the use of colors is the same given *above*. Contact authors for color figures.

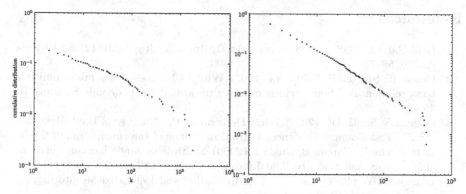

Fig. 6. Banks net: degree distribution and degree correlation.

plotted in the Fig. 6. Both of them are fat-tailed distributions, reproducing the heterogeneous behavior of banks.

7 Discussion and Conclusions

In this paper we have analyzed the credit relationships between firms and banks in Italy.

We observe that small banks have a smaller number of contracts and large banks a big number of contracts. Moreover big firms tend to receive loans from big banks but also from some small banks; on the contrary, small firms receives loans from small (local) banks. In general big firms prefer multiple relationships, while small firms tend to a single relationship. This can be explained considering the exposure to risk: in fact the firms with single linkage are in general less solvent [13]. How do the different features of firms influence the topological characteristics of the networks is object of study of a second paper [13]. In multiple relationships, the heterogeneity between banks cofinancing the same firm emerges.

The topology of the underlying credit network plays a crucial role in bankruptcy diffusion. Further analysis will focus on domino effects of failures in the observed topologies of networks [31].

It should be interesting to study the relationship between rate and firm performance indicators (as solvency, added value). Moreover information about the geographical location of firms and banks would allow to study contracts based on personal trust toward debtors.

A comparison between the network structure of bank-firm credit networks of many countries may permit to understand how different institutional rules produce different emerging structures [32, 33].

Acknowledgement. We acknowledge the Eurace project, financing this research. We are grateful to Stefano Battiston for the preparation of dataset.

References

1. Delli Gatti D, Gallegati M, Greenwald B, Russo A, Stiglitz JE (2006) Physica A, 370: 68–74
2. Ogawa K, Sterken E, Tokutsu I (2007) Why do Japanese firms prefer multiple bank relationship? Some evidence from firm-level data, Economic Systems 31: 49–70
3. Ongena S, Smith DC (2000) What Determines the Number of Bank Relationships? Cross-Country Evidence, Journal of Financial Intermediation, 9: 26–56
4. Detragiache E, Garella P, Guiso L (2000) Multiple vs single banking relationships: theory and evidence, The Journal of Finance LV(3)
5. Diamond DW (1984) Financial Intermediation and Delegated Monitoring, review of economic Studies, 51: 393
6. Agarwal R, Elston JA (2001) Bank-firm relationships, financing and firm performance in Germany, Economics Letters, 72:225–232
7. Farinha LA, Santos JAC (2000) Switching fro single to multiple bank lending relationships: determinants and implications, Bank for International Settlement, Working paper 83
8. Forestieri G, Tirri V, Rapporto banca-impresa. Struttura del mercato e politiche del prezzo, Ente Luigi Einaudi, Quaderni di Ricerche 31
9. Stanca L, Arosio S, Bonanno V, Cavalieri C (2005) La struttura finanziaria in Italia: evoluzione del rapporto banche e imprese, Politica Economica
10. Pagano M, Panetta F, Zingales L (1998) Why Do Companies Go Public? An Empirical Analysis, The Journal of Finance 53: 27–64
11. www.bvdep.com
12. Annual report on vigilance activites of Bank of Italy, May 2005
13. De Masi G, Gallegati M (2007) A detailed analysis of Italian banks-firms lending network and its evolution, working paper
14. The Italian Institute for Statistics provides a classification of Italian firms in micro-firms, small-size, medium-size and large. The classification depends on the numeber of workers: 1) less than 10 workers, micro; 2) between 10 and 50 small; 3) between 50 and 250 medium; 4) more than 250 large.
15. Salza E (2004) Il sistema produttivo e il rapporto banca-impresa, Impresa e Stato, 67
16. Dorogovtsev SN, Mendes JFF (2003) Evolution of Networks From Biological Nets to the Internet and the WWW, Oxford University Press
17. Caldarelli G (2007) Scale-Free Networks. Complex Webs in Nature and Technology, Oxford University Press
18. Gallegati M, Kirman A (1999) Beyond the representative agent, Edward Elgar Publishing
19. Serrano MA, Boguñá M (2003) Topology of the world trade web, Physical Review E 68: 015101
20. Garlaschelli D, Loffredo MI (2004) Fitness-Dependent Topological Properties of the World Trade Web, Physical Review Letters 93: 188701
21. Iori G, De Masi G, Precup O, Gabbi G, Caldarelli G (2007) A network analysis of the Italian overnight money market, to appear in Journal of Economic Dynamics and Control
22. Garlaschelli D, Battiston S, Castri M, Servedio VDP, Caldarelli G (2005) The scale free topology of market investments, Physica A 350: 491–499

23. Reichardt J, Bornholdt S (2005) eBay users form stable groups of common interest, pre-print: physics/0503138
24. Peltomaki M, Alava M (2006) Correlations in bipartite collaboration networks, J. Stat. Mech., P01010
25. Sneppen K, Rosvall M, Trusina A (2004) Europhys. Lett., 67: 349–354
26. Guillaume JL, Latapy M (2004) Bipartite structure of all complex networks, Inf. Process. Lett., 90 (5)
27. Souma W, Fujiwara Y, Aoyama H (2003) Complex Networks and Economics, Physica A 324: 396–401
28. Lind PG, Herrmann HJ (2007) New approaches to model and study social networks, arxiv:physics/0701107
29. Strogatz SH (2001) Exploring complex networks, Nature 410: 268
30. Newman MEJ (2001) Scientific collaboration networks. II. Shortest paths, weighted networks, and centrality, Phys. Rev. E 64: 026118
31. De Masi G, Gallegati M, Greenwald B, Stiglitz JE, Economic stability in debt/credit networks, working paper
32. De Masi G, Fujiwara Y, Gallegati M, Greenwald B, Stiglitz JE, Short-term and long-term credit relationships in Japanese Economy, working paper
33. De MasiG, Battiston S, Gallegati M, Comparison of network topologies of debt-credit networks in European countries, working paper

Econophysicists Collaboration Networks: Empirical Studies and Evolutionary Model

Menghui Li[1], Jinshan Wu[2], Ying Fan[1*], and Zengru Di[1]

[1] Department of Systems Science, School of Management,
Beijing Normal University, Beijing 100875, P.R.China
limh@mail.bnu.edu.cn, yfan@bnu.edu.cn(author for correspondence),
zdi@bnu.edu.cn
[2] Department of Physics & Astronomy,
University of British Columbia, Vancouver, B.C. Canada, V6T 1Z1
jinshanw@phas.ubc.ca

Scientific collaboration network gives a nice description for communications among scientists. In order to show the status of the research in Econophysics, we have collected papers in Econophysics and constructed a network of scientific communication to integrate idea transportation among econophysicists by collaboration, citation and personal discussion. Inspired by scientific collaboration networks, especially our empirical analysis of econophysicists network, an evolutionary model for weighted networks is proposed. The model shows the scale-free phenomena in degree and vertex weight distribution. The results of short term evolution are consistent well qualitatively with the empirical results.

1 Introduction

Now the scientific collaboration networks(SCN) is of general interest for understanding the topological of complex networks. Several fields of collaboration networks are investigated in recent years, for example, biomedical, theoretical physics, high-energy physics and computer science [1,2], mathematics and neuro-science [3,4], condensed matter physics [5] and so on.

Econophysics is a new field developed recently by the cooperation between economists, mathematicians and physicists. More and more researchers in finance take up statistical physics to explore the dynamical and statistical properties of financial data, including time series of stock prices, exchange rate, size of organizations and so on. And also more and more physicists from statistical physics and complexity turn to working in finance, as an important and copies research subject. It is interesting to investigate the corresponding research development in this new born scientific research field, to know the work status and to understand the idea transportation among scientists.

In scientific collaboration networks, there are more than one level interactions on a link, such as collaboration, citation and personal discussion. They are the ways of idea transportation but with different contributions. When we analyze this transportation as a whole, we have to use different weights to measure these different contributions. Also, even for the same level interaction, such as collaboration, not only the existence of connection but the times of collaboration is valuable information. So to fully characterize the interactions in real networks, link weight, which integrating the information of times and levels, should be taken into account.

Actually, how to assign the weight into link and how will it evolve are the key problems both for empirical and modeling analysis. The way to measure the weight for weighted networks has been introduced differently by some authors, such as the number of flights or seats between any two cities in airport networks [5–7], the reaction rate in metabolic networks [8], the happening number of the event in scientific collaboration networks [2, 5] and so on. In the work of modeling weighted networks, there are also several ways to assign weight to link. Some previous approaches have transferred some quantities from non-weighted networks into the weight of links [9, 10]. In some models, weights on links are generated from some a priori distribution [11–13]. The link weight in above models does not change with the evolution of the networks. Some other models [14–17] couple the link weight with network evolution. But all of them require some extra mechanism for the evolution of link weights.

Our empirical investigation on SCN try to integrate different kinds and different strength of interactions into account. It also gives us some hints on modeling weighted scientific collaboration networks. As we mentioned before, link weight in collaboration networks is usually related closely to the number of connections. In our model, we keep the relationship between weight and the number of connections. Only the quantities directly rooted in networks are used. So the picture of the evolution looks like the number of connections evolves according to weight, and then the new connection number comes into the weight, which drives the evolution of the network again. Another important improvement of our model is the introduction of distance-dependent preferential attachment(DDPA), i.e., the δ term in our model. It is helpful to increase the clustering coefficient of networks. The DDPA mechanism requires only the information about the local structure but it improves the situation about clustering coefficient obviously. Furthermore, the models constructed in this way do not require any extra dynamical process of weight.

2 Empirical Networks and Statistical Results

Concentrating on main topics of Econophysics, we have collected papers from the three web sites listed below. (1) Econophysics home page at

http://www.econophysics.org. (2)Econophysics at Physica A at http://www.
elsevier.com/locate.econophys. (3) ISI at http://isi.knowlege.com. We collected publish information of papers published from 1992 to 7/30/2004, including 808 papers and totally 819 authors. We believe that many papers in
Econophysics are not collected into this database for some reasons.

Based on the data set, we extracted the times of three relations between
every two scientists to form a file of data recorded as 'S_1 S_2 x y z', which
means author S_1 has collaborated with author S_2 'x' times, cited 'y' times of
S_2' papers and thanked S_2 'z' times in all S_1's acknowledgements. One can
regard this record as data of three different networks, but from the idea of
transportation and development of this field, it's better to integrate all these
relations into a single one by the weight of connection. Here we have to mention
that in order to keep our data set closed, we only count the cited papers
collected by our data set and just select the people in acknowledgements
which are authors in our data set. We convert the times to weight by

$$w_{ij} = \sum_{\mu} w_{ij}^{\mu},\qquad(1)$$

in which μ can only take a value from $1, 2, 3$. So w_{ij}^{μ} is one of the three
relationships–coauthor, citation or acknowledgement and is defined as

$$w_{ij}^{\mu} = \tanh\left(\alpha_{\mu}T_{ij}^{\mu}\right),\qquad(2)$$

where T_{ij}^{μ} is the time of μ relationship between i and j.

We suppose that the weight should not increase linearly, and it must reach
a limitation when the times exceeds some value. So we use tanh function to
describe this nonlinear effect. We also assume that the contributions to the
weight from these three relations are different and they can be represented by
the different values of α_{μ}. $0.7, 0.2, 0.1$ are used for $\alpha_1, \alpha_2, \alpha_3$ in this paper.

The similarity is used here as the weight, after the network has been constructed, it is converted into dissimilarity weight as

$$\tilde{w}_{ij} = 3/w_{ij}\quad(if\ w_{ij} \neq 0).\qquad(3)$$

It's timed by 3 because the similarity weight $w_{ij} \in [0, 3]$. Hence we have
$\tilde{w}_{ij} \in [1, \infty]$, and it is corresponding to the "distance" between vertices. All
quantities are calculated under dissimilarity weight if not mentioned.

Since this scientists network is directed, we have three definitions of degree:
out degree, in degree and total degree. And also three ones for vertex weight.
Figure 1 shows all these distributions. In addition, We have defined another
interesting quantity - weight per degree, which is the quotient of weight and
degree. It also has three value, in, out and total. Figure 2 shows the distribution of total one. Weight per degree is one of the inherent variables of
vertex. It has some relations with the working style of scientists. Preferring
to cooperate widely not deeply may lead to large weight per degree, while for

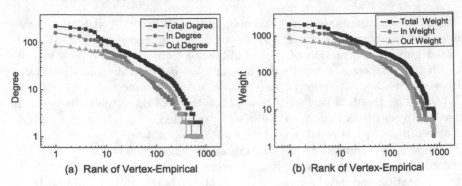

Fig. 1. Zipf plot of degree (**a**) and vertex weight (**b**) distribution.

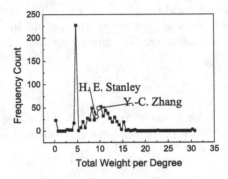

Fig. 2. Distribution of weight per degree.

other scientists maybe wish to cooperate with few people but more deeply and this result in small weight per degree. Maybe this quantity can be used as a measure of the working style of a scientist(see details in papers [18]).

Other important geometrical quantities for the network are the vertex and link betweenness [19]. Here in Fig. 3, we give vertex betweenness and link betweenness distributions. Betweenness describes the relative importance of

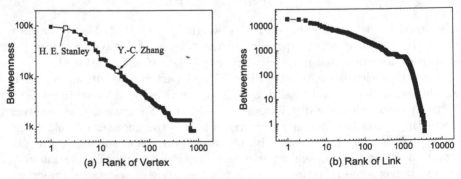

Fig. 3. Zipf plot of vertex (**a**) and link (**b**) betweenness distribution.

a vertex or link in the communication in the network. A great betweenness implies it is a key point for global communication.

Besides above quantities, community structure is also an important properties of networks. It can give us deeper understanding to the properties of network. Here we take the largest connected group of Econophysicists co-authorship network(neglecting citation and acknowledgement) as our research subject. It includes 271 nodes and 371 edges. After analyzing the econophysicists collaboration network by GN algorithm [19] based on above definition of link weight, we draw the community structure of Econophysicists network (as shown in Fig. 4). The best division of GN algorithm shows that the scientists, who are in the same university, institute or interested in similar research topic, are clustered to one community. For example, in Fig. 4, members of no. 1 community are most from Boston University, USA. The members of no. 2 community are most Tokyo, Japan. Members of no. 4 community are mainly from the University of Southern California and the University of California at Los Angeles. There are also other communities which members are interested in the same topic, as no. 9 community whose members from different areas but focus on financial market(see details in paper [20]).

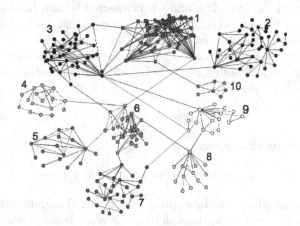

Fig. 4. Community structure of Econophysicists co-authorship network.

3 Models and Theoretical Analysis

3.1 The Model

An N-vertex weighted network is defined by an $N \times N$ matrix W. Its element w_{ij} represents the weight on link from vertex i to j. "Similarity weight" is used here. So the larger is the weight, the closer is the relation between the two end vertices. Suppose the link weight w_{ij} is related to the connection number T_{ij} between vertex i and j by

$$w_{ij} = f(T_{ij}). \tag{4}$$

Our model is given as follow. Starting from a fully connected network with n_0 vertices and initial connecting times $T_{ij} = 1$ (then initial weight $w_{ij} = f(1)$), at every time step,

1. One new vertex is added into this network, and l old vertices are randomly chosen from the existing network.
2. Every one (denoted as vertex n) of these $1+l$ vertices can initially activate a temptation to build up m connections. The probability for link from n connecting onto vertex i is given by

$$\Pi_{n \to i} = (1-p)\frac{k_i}{\sum_j k_j} + (p-\delta)\frac{w_i}{\sum_j w_j} + \delta\frac{l_{ni}}{\sum_{j \in \partial_n^d} l_{nj}}, \tag{5}$$

where k_i is the degree of vertex i, $w_i = \sum_j w_{ji}$ is the "onto" vertex weight or the strength of vertex i, l_{ni} is the similar distance from n to i. If n and i is connected, $l_{ni} = w_{ji}$. If n and i is connected by two links (with weight w_{ns} and w_{si} respectively), $l_{ni} = w_{ns}w_{si}/(w_{ns} + w_{si})$. ∂_n^d is the set of all n's neighbors to the dth nearest.
3. After we get an end vertex i^* chosen from all vertices over the existing network by the probability above in Eq. (5), the connecting times between vertex n and i^* becomes

$$T_{ni^*}(t+1) = T_{ni^*}(t) + 1. \tag{6}$$

4. The link weight changes as

$$w_{ni^*}(t+1) = f(T_{ni^*}(t+1)). \tag{7}$$

The model defined above can be applied to undirected and directed networks. The expanded directed network model for the comparison with empirical results is given in Sect. 4. In the following analysis we assume that $w_{ij} = w_{ji}$ and the linear function is used for the relationship between connecting times and weight for the simplicity,

$$w_{ij} = \alpha T_{ij}. \tag{8}$$

3.2 Analytic Results of the Weight Distribution

In this subsection, we focus on the behavior of the simplest pure weight-driven model, that is $p = 1$ and $\delta = 0$ in Eq. (5). For the link weight given by Eq. (8) with $\alpha=1$, the weight of vertex is given by $w_i = \sum_j T_{ji}$. So the connection probability is

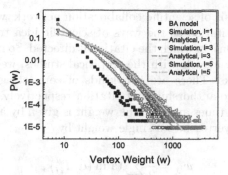

Fig. 5. For a pure weight-driven model, analytical results of Eq. (11) are compared with the computer simulations under different l.

$$\Pi_{n \to i} = \frac{w_i}{\sum_j w_j} \tag{9}$$

The rate equation for the evolution of the average number of vertices with weight w at time t is

$$N(w, t+1) = N(w,t) + m \cdot (1+l) \cdot \frac{(w-1) \cdot N(w-1, t) - w \cdot N(w,t)}{\sum_w w \cdot N(w,t)}$$
$$- \frac{l \cdot N(w,t)}{N} + \frac{l \cdot N(w-m, t)}{N} + \delta_{w,m} \tag{10}$$

Here $\sum_w w \cdot N(w, t-1) = 2 \cdot E_0 + 2 \cdot m \cdot (1+l) \cdot t$ is the total weight and $N = n_0 + t$ is the size of network at time t. When t is larger($t \gg 1$) enough, we get the asymptotic distribution of vertex weight,

$$p(w) \propto (w + 2 \cdot l \cdot m)^{-3}. \tag{11}$$

Comparison of the analytical results of Eq. (11) with computer simulations, Fig. 5 shows that they are very consistent with each other. We can find that the lower end of vertex weight is obviously affected by the parameter l, and departs from a power-law, while the upper end is still distributed as a power-law. The distribution of degree and link weight also obey power law. The correlation of vertex weight and degree is nonlinear, namely $w \sim k^\eta$. When $\delta \neq 0$, the clustering coefficient reaches a stable value after a period of evolution, which is much larger than that of BA model (nearly 0), and even comparable with empirical results(see details in paper [21]). All of above results are consistent with the typical results from empirical studies qualitatively.

4 Extended Model and Comparison with Empirical Results

In the real world, relations of vertices usually have multi-levels. Different level relations usually have different contributions to link weight. For instance, in the empirical analysis of scientific collaboration network, we considered both co-authorship and citation as the ways of scientific idea transportation with different contributions, in which the citation is directed. So in order to compare the results from our models with the empirical studies, we extend our model to directed networks. There are two kinds of connecting times T_{ij}^{μ}, where $\mu = 1, 2$ refers to co-authorship and citation respectively. Here the relation between connecting times and the link weight is given by a tanh function. So the two T_{ij}^{μ} are converted into a single weight by

$$w_{ij} = \frac{1}{2} \sum_{\mu} \tanh \left(\alpha_{\mu} T_{ij}^{\mu} \right), \tag{12}$$

where w_{ij} is normalized to 1. And the probability distribution to chose the end vertex is consistently transformed as

$$\Pi_{n \to i} = \sum_{\mu} p^{\mu} \frac{w_i}{\sum_j w_j}, \tag{13}$$

while $\sum_{\mu} p^{\mu} = 1$. After vertex i^* is chosen as the end vertex of a relation μ between n and i^* according to above probability distribution, the connecting times evolves as

$$T_{ni^*}^{\mu} (t + 1) = T_{ni^*}^{\mu} (t) + 1. \tag{14}$$

For $\mu = 1$, after that we need to set $T_{i^*n}^1 (t + 1) = T_{ni^*}^1 (t + 1)$. For $\mu = 2$, we skip this step. Then these two counts are integrated into the link weight by definition Eq. (12). From the simulation results, we can find that the final steady distributions of the total, in, and out degree and weight all obey power law, especially for the upper end.

We compare the results of the model with empirical results from Econophysicists collaboration networks. The distribution of quantities numerical simulations, such as degree, vertex weight(Fig. 6), vertex betweenness and

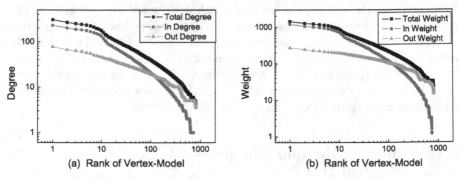

Fig. 6. Zipf plot of degree (**a**) and vertex weight (**b**). The parameters are $n_0 = 10, m = 5, l = 1, p^1 = 0.2, p^2 = 0.8, \alpha^1 = 0.7, \alpha^2 = 0.3, N = 819$.

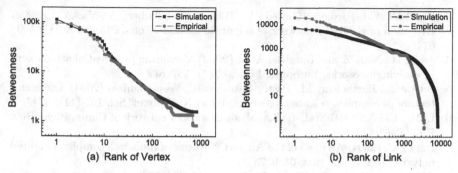

Fig. 7. Zipf plot of vertex betweenness (**a**) and link betweenness (**b**) for empirical studies and simulations.

link betweenness(Fig. 7) are consistent well with that of Econophysicists networks(Fig. 1 and Fig. 3) qualitatively. It seems that the model reveals some basic mechanisms of the evolution of collaboration network.

5 Conclusions

In summary, we have constructed a small network by collecting papers in Econophysics. A new definition of weight and some fundamental properties are analyzed. The idea to integrate networks with multilevel interactions and different strength may be helpful for further studies on weighted networks.

Inspired by scientific collaboration networks, a simple evolutionary model for weighted network is presented. A natural way for link weight evolution is introduced. The number of connections between any two vertices is converted to the link weight. So the link weight evolves naturally with the growth of networks and then drives the evolution of network again. The model shows a good consistence with empirical results of Econophysicists collaboration network. This reveals that the model has captured some basic mechanisms for the evolution of scientific collaboration networks.

The work is partially supported by 985 project and NSFC under the grant Nos. 70431002 and 70471080.

References

1. Newman M.E.J. (2001) The structure of scientific collaboration networks, Proc. Natl. Acad. Sci. USA 98: 404–409
2. Newman M.E.J. (2001) Scientific collaboration networks. I. Network construction and fundamental results, Phys. Rev. E 64: 016131; Newman M.E.J.(2001) Scientific collaboration networks. II. Shortest paths, weighted networks, and centrality, Phys. Rev. E 64: 016132

3. Barabási A.-L., Jeonga H., Neda Z., Ravasz E., Schubert A., Vicsek T.(2002) Evolution of the social network of scientific collaborations, Physica A 311: 590–614

4. Jeong H., Néda Z. and Barabási A. L. (2003) Measuring preferential attachment in evolving networks, Europhys. Lett., 61(4): 567–572

5. Barrat A., Barthélemy M., Pastor-Satorras R., Vespignani A. (2004) The architecture of complex weighted networks, Proc. Natl. Acad. Sci. 101 (11): 3747

6. Li W., Cai X.(2004) Statistical analysis of airport network of China, Phys. Rev. E 69: 046106

7. Bagler G. (2004) Analysis of the Airport Network of India as a complex weighted network, arXiv:cond-mat/0409773

8. Almaas E., Kovacs B., Vicsek T., BarabásiA.-L.(2004) Global organization of metabolic fluxes in the bacterium Escherichia coli, Nature 427: 839

9. Yook S.H., Jeong H., Barabási A.-L. (2001) Weighted evolving networks, Phys. Rev. Lett. 86: 5835–5838

10. Zheng D., Trimper S., Zheng B., Hui P.M. (2003) Weighted scale-free networks with stochastic weight assignments, Phys. Rev. E 67: 040102

11. Antal T., Krapivsky P.L., Weight-driven growing networks, Phys. Rev. E 71: 026103

12. Goh K-I, Noh J D, Kahng B and Kim D (2005) Load distribution in weighted complex networks, Phys. Rev. E 72: 017102

13. Park K., Lai Y.-C., Ye N. (2004) Characterization of weighted complex networks, Phys. Rev. E 70: 026109

14. Barrat A, Barthélemy M and Vespignani A (2004) Weighted Evolving Networks: Coupling Topology andWeight Dynamics, Phys. Rev. Lett. 92: 228701

15. Bianconi G (2005) Emergence of weight-topology correlations in complex scale-free networks, Europhysics letters 71: 1029–1035

16. Dorogovtsev S N and Mendes J F F (2004) Minimal models of weighted scale-free networks, arXiv: cond-mat/0408343

17. Wang W., Wang B., Hu B., Yan G., Ou Q.(2005) General dynamics of topology and traffic on weighted technological networks, Phys. Rev. Lett. 94: 188702

18. Fan Y., Li M., Chen J., Gao L., Di Z., Wu J. (2004) Netwok of econophysicists : a weighted network to investigate the development of ecnophysics, International Journal of Modern Physics B Vol. 18, Nos. 17–19: 2505–2511; Li M., Fan Y., Chen J., Gao L., Di Z., Wu J. (2005) Weighted networks of scientific communication: the measurement and topological role of weight, Physica A 350: 643–656

19. Girvan M. and Newman M. E. J. (2002) Community structure in social and biological networks, Proc. Natl. Acad. Sci. USA, 99: 7821

20. Zhang P., Li M., Wu J., Di Z., Fan Y. (2006) The analysis and dissimilarity comparison of community structure, Physica A 367: 577–585

21. Li M., Wang D., Fan Y., Di Z. and Wu J. (2006) Modelling weighted networks using connection count, New Journal of Physics 8: 72; Li M., Wu J., Wang D., Zhou T., Di Z., Fan Y. (2007) Evolving model of weighted networks inspired by scientific collaboration networks, Physica A 375: 355–364

Part III

Income, Stock
and Other Market Models

The Macro Model of the Inequality Process and The Surging Relative Frequency of Large Wage Incomes

John Angle

Inequality Process Institute,
Post Office Box 429, Cabin John, Maryland, 20818, USA
angle@inequalityprocess.org

Revision and extension of a paper, 'U.S. wage income since 1961: the perceived inequality trend', presented to the annual meetings of the Population Association of America, March-April 2005, Philadelphia, Pennsylvania, USA. On-line at: http://paa2005.princeton.edu/download.aspx?submissionID=50379.

1 The Surge In Wage Income Nouveaux Riches in the U.S., 1961–2003

Angle (2006) shows that the macro model of the Inequality Process provides a parsimonious fit to the U.S. wage income distribution conditioned on education, 1961–2001. Such a model should also account for all time-series of scalar statistics of annual wage income. The present paper examines one such time-series, the relative frequency of large wage incomes 1961–2003. Figure 1 shows an aspect of this kind of statistic: the larger the wage income, the greater the proportional increase in its relative frequency. The phenomenon that Fig. 1 shows, a surge in wage income nouveaux riches, has caused some alarm and given rise to fanciful theories. The present paper shows that the macro model of the Inequality Process (IP) accounts for this phenomenon. In fact, it is simply an aspect of the way wage income distributions change when their mean and all their percentiles increase, which they do simultaneously, i.e., it is good news for a much larger population than the nouveaux riches alone. Nevertheless, many economists and sociologists have interpreted the surge in wage income nouveaux riches[1] as an alarming bifurcation of the U.S. wage income distribution into two distributions, one poor, the other rich, a 'hollowed out'

---

[1] The term nouveaux riches perhaps brings to mind the new wealth of entrepreneurs most of whose income is from tangible assets. Nouveaux riches is used here only to name the earners of wage income who have begun to earn a wage income much larger than the average.

distribution. Fear of the 'hollowing out' of the U.S. wage income distribution has not only roiled academia but has resulted in alarmed press reports and even become an issue in the 2004 U.S. presidential campaign.

The present paper shows that an increase in mean wage income decreases the relative frequency of wage incomes smaller than the mean and increases the relative frequency of wage incomes greater than the mean. Distance from the mean of a particular wage income, call it x_0, is a factor in how fast the relative frequency of wage incomes of that size change. For x_0's greater than the mean, the greater x_0, the greater the proportional growth in its relative frequency. There is an analogous and compensating decrease in the relative frequency of wage incomes smaller than the mean. The IP's macro model implies that the wage income distribution stretches to the right when the unconditional mean of wage income increases, explaining both the surge in wage income nouveaux riches and the fact that the bigger wage income percentile has grown more than the smaller wage income percentile. Data on U.S. wage incomes 1961–2003 confirm the implications of the IP's macro model.

The data on which this paper is based are the pooled cross-sectional time series formed from 'public use samples' of the records of individual respondents to the March Current Population Surveys (CPS) 1962–2004. The March CPS is a survey of a large sample of U.S. households conducted by the U.S. Bureau of the Census.[2] Each March CPS asks questions about the level of education of members of the household and their sources of income in the previous calendar year. The study population is U.S. residents with at least $1 (one U.S. dollar) in annual wage income who are at least 25 years old. All dollar amounts in this paper are in terms of constant 2003 dollars, i.e., adjusted for changes in the purchasing power of a dollar from 1961 through 2003. See Appendix A.

Figure 1 illustrates one of two related ways to measure the surge in wage income nouveaux riches in the U.S. 1961–2003. Figure 1 measures change in the relative frequencies in the far right tail of the wage income distribution, the distribution of wage income recipients over the largest wage incomes. This dynamic of the distribution can be readily understood in terms of the algebra of the macro model of the Inequality Process. The other way to measure the surge is via change in percentiles of wage incomes (in constant dollars), e.g., the 90th percentile. Figure 2 shows the 90th percentile of wage incomes increasing more in constant dollars than the 10th percentile, a small wage

---

[2] These data have been cleaned, documented and made readily accessible as a user-friendly database by Unicon Research Corporation (2004), a public data reseller supported by the (U.S.) National Institutes of Health. The author welcomes replication of tests of the Inequality Process either with the data used in this paper, available at nominal cost from the Unicon Research Corporation, 111 University Drive, East, Suite 209, College Station, Texas 77840, USA, or replication with comparable data from additional countries.

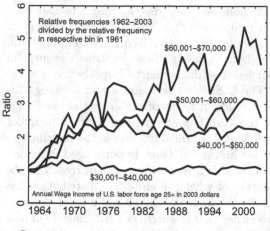

Fig. 1. Ratio of relative frequencies 1962–2003 in wage income bins $1–$10,000, $10,001–$20,000 etc. in terms of constant 2003 dollars to the relative frequency in each bin in 1961 Source: Author's estimates of data of the March Current Population Survey.

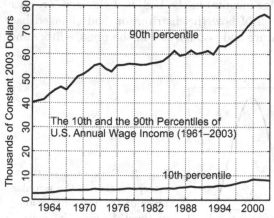

Fig. 2. Source: Author's estimates from March CPS data.

income. This second way of measuring the surge, as a stretching of the distribution to the right, is implied by the IP's macro model but requires numerical integration to demonstrate.

The surge in wage income <u>nouveaux riches</u> in the U.S. 1961–2003 has been a focus of concern in U.S. labor economics and sociology journals. A substantial fraction of contributions to this literature have interpreted the surge as part of the transformation of the U.S. wage income into a bimodal, U-shaped distribution. See Fig. 3 for a caricature of this thesis and the 'hollowing out' literature itself (Kuttner, 1983; Thurow, 1984; Lawrence, 1984; Blackburn and Bloom, 1985; Bradbury, 1986; Horrigan and Haugen, 1988; Coder, Rainwater, and Smeeding, 1989; Levy and Michel, 1991; Duncan, Smeeding and Rodgers, 1993; Morris, Bernhardt, and Handcock, 1994; Wolfson, 1994; Esteban and Ray, 1994; Jenkins, 1995; Beach, Chaykowski, and Slotsve, 1997; Wolfson, 1997; Burkhauser, Crews Cutts, Daly, and Jenkins, 1999; Esteban and Ray, 1999; Duclos, Esteban and Ray, 2004;). The emergence of a U-shaped wage income distribution has been termed the 'hollowing out' or 'polarization' of the wage income distribution. A 'hollowed out' distribution has also been called a 'barbell distribution'. In a 'hollowing out' the relative frequency of middling wage incomes decreases while the relative frequencies of small and large wage incomes increase. The 'hollowing out' thesis explains the surge in large wage incomes but it is burdened with having to hypothesize a surge in small wage incomes as well. The U.S. labor economics and sociology literatures on wage income measure trends in terms of scalar statistics of wage income, mostly the median plus the grab bag of statistics referred to under the rubric 'statistics of inequality'. Models of the dynamics of the distribution are rare and never prominent in this literature. Thus, hypothesized dynamics of the distribution have been used in this literature to explain trends in the scalar statistics without confirmation of what the empirical distribution has actually been doing. The rise of the thesis of the 'hollowing out' of the U.S. wage income distribution requires either that researchers were unaware of how the empirical distribution of wage incomes in the U.S. had changed, or, once the 'hollowing out' interpretation of how the distribution had changed had become established in the literature and popularized in the press, editors and

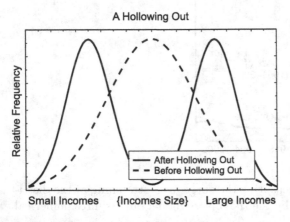

Fig. 3. Caricature of the interpretation in the popular press of the transformation of the U.S. wage income distribution into a bimodal, 'barbell' distribution in recent decades. See Appendix B.

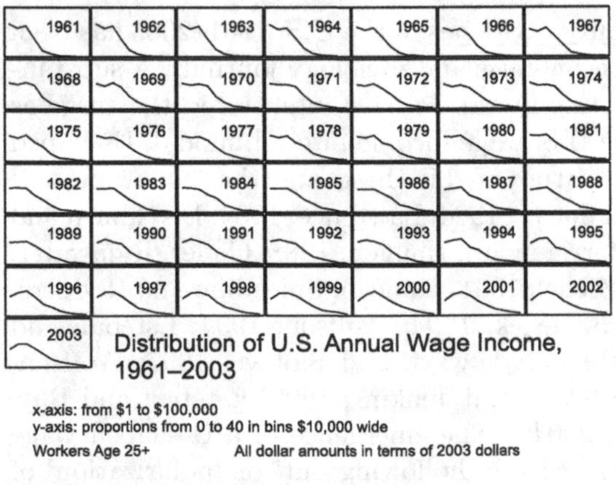

Fig. 4. Source: Author's estimates from March CPS data.

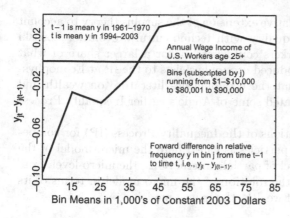

Fig. 5. Forward difference between mean relative frequencies in 1961–1965 $(t-1)$ and mean relative frequencies in 1999–2003 (t). Source: Author's estimates from data of the March CPS.

reviewers were unable to accept evidence to the contrary, i.e., the journals that established the 'hollowing out' thesis could not correct their error.[3]

Figure 4 displays estimates of the U.S. wage income distribution from 1961 through 2003. It is clear that, indeed, its right tail thickened over these 43 years, i.e., the relative frequency of large wage incomes increased validating half of the 'hollowing out' hypothesis. However, it is as clear in Fig. 4 that the left tail of the distribution of wage incomes, the distribution of workers over small wage incomes, thinned, that is, the relative frequency of small wage incomes decreased. The 'hollowing out' thesis requires both to increase simultaneously.

Figure 5 displays the forward differences between mean relative frequencies of wage income between two ten year periods, 1961–1970 and 1994–2003. Ten year means are taken to smooth the relative frequencies. Figure 5 shows that the relative frequency of small wage incomes fell between these two periods and larger wage incomes increased, just as an inspection of Fig. 4 would lead one to believe.

2 The Micro- and Macro-Models of the Inequality Process (IP)

The Inequality Process (IP) is a stochastic binary interacting particle system derived from an old verbal theory of economic anthropology (Angle, 1983 to 2006). People are the particles. Wealth is the positive quantity transferred between particles. The theory from which the IP is derived is the 'surplus theory of social stratification'. It asserts that economic inequality among people originates in competition for surplus, societal net product. The IP literature dates from (Angle, 1983). The IP is an abstraction of a mathematical model from

---

[3] The author tried to correct this literature in terms familiar to its contributors over a period of many years but, so far, has been unable to publish in any of the journals responsible for popularizing the 'hollowing out' thesis.

Gerhard Lenski's (1966) speculative extension of the surplus theory to account for the decrease in wealth inequality with techno-cultural evolution. Lenski thought that more skilled workers could bargain for a larger share of what they produce. Lux (2005) introduced econophysicists to the IP at Econophys-Kolkata I and pointed out that the econophysics literature on wealth and income distribution had replicated some of Angle's earlier Inequality Process findings.

The pair of transition equations of the Inequality Process (IP) for competition for wealth between two particles is called here the micro model of the IP to distinguish the IP's model of particle interactions, the micro-level, from the model that approximates the solution of the micro model in terms of its parameters, the distribution of wealth, the macro model.

2.1 The Micro-Model of the Inequality Process (IP)

Since the macro-model of the Inequality Process (IP) is derived from the IP's micro model, description of the former should begin with description of the latter. Consider a population of particles in which particles have two characteristics. Wealth is one such trait. Particle i's wealth at time t is denoted, x_{it}. Wealth is a positive quantity that particles compete for pairwise in zero-sum fashion, i.e., the sum of wealth of two particles after their encounter equals the sum before their encounter. The other particle characteristic is the proportion of wealth each loses when it loses an encounter. That proportion is the particle's parameter u. So, from the point of view of a particular particle, say particle i, the proportion of wealth it loses, if it loses, ω_i, is predetermined. When it wins, what it wins is, from its point of view, a random amount of wealth. Thus there is an asymmetry between gain and loss from the point of view of particle i. Long term each particle wins nearly 50% of its encounters. Wealth is transferred, long term, to particles that lose less when they lose, the robust losers.

Let particle i be in the class of particles that loses an ω_ψ fraction of their wealth when they lose, i.e., $\omega_i = \omega_\psi$. In the IP's meta-theory, Lenski's speculation, workers who are more skilled retain a larger proportion of the wealth they produce. So smaller ω_ψ in the IP's meta-theory represents the more skilled worker. Worker skill is operationalized in tests of the IP, as is usual in labor economics, by the worker's level of education, a characteristic readily measured in surveys. For tests of the IP on wage income data by level of worker education, the IP's population of particles is partitioned into equivalence classes of its particles' ω_ψ by the corresponding level of education, so that the proportion formed by workers at the ψth level of education of the whole labor force, u_ψ, ('u' to suggest 'weight'), is the proportion formed by the ω_ψ equivalence class of the population of particles. The u_ψ's are estimated from data and so are the ω_ψ's by fitting the comparable statistic of the IP to either micro-level data (the dynamics of individual wage incomes) or macro-level data (the dynamics of wage income distributions). The IP's meta-theory implies that estimated ω_ψ's should scale inversely with worker education level, based on the assumption that the more educated worker is the more productive worker. Nothing in the testing of this prediction forces this outcome. The predicted outcome holds in U.S. data on wage incomes by level of education as demonstrated in Angle (2006) for 1961–2001 and below for 1961–2003. While there is no apparent reason why this finding should not generalize to all industrial labor forces in market economies, the universality of the prediction is not yet established. The IP is a highly constrained model. Its predictions are readily falsified if not descriptive of the data.

The transition equations of the competitive encounter for wealth between two particles in the Inequality Process' (IP's) micro model are:

$$x_{it} = x_{i(t-1)} + \omega_{\theta j} d_{it} x_{j(t-1)} - \omega_{\psi i}(1 - d_{it})x_{i(t-1)}$$
$$x_{jt} = x_{j(t-1)} - \omega_{\theta j} d_{it} x_{j(t-1)} + \omega_{\psi i}(1 - d_{it})x_{i(t-1)} \qquad (1)$$

where $x_{i(t-1)}$ is particle i's wealth at time $t - 1$,

$$d_{it} = \begin{cases} 1 & \text{with probability .5 at time } t \\ 0 & \text{otherwise.} \end{cases}$$

and,

$\omega_{\psi i}$ = proportion of wealth lost by particle i when it loses (the subscript indicates that particle i has a parameter whose value is $\omega_{\psi i}$; there is no implication that the ω_ψ equivalence class of particles has only one member or that necessarily $\omega_{\psi i} \neq \omega_{\theta j}$);

$\omega_{\theta j}$ = proportion of wealth lost by particle j when it loses.

Particles are randomly paired; a winner is chosen via a discrete 0,1 uniform random variable; the loser gives up a fixed proportion of its wealth to the winner. In words, the process is:

Randomly pair particles. One of these pairs is particle i and particle j. A fair coin is tossed and called. If particle i wins, it receives an ω_θ share of particle j's wealth. If particle j wins, it receives an ω_ψ share of particle i's wealth. The other particle encounters are analogous. Repeat.

The asymmetry of gain and loss is apparent in Fig. 6, the graph of forward differences, $x_{it} - x_{i(t-1)}$ against wealth, $x_{i(t-1)}$, resulting from (1). The Inequality Process differs from the Saved Wealth Model, a modification of the stochastic model of the Kinetic Theory of Gases that generates a gammoidal stationary distribution discussed by Chakraborti, Chakrabarti (2000); Chatterjee, Chakrabarti, and Manna (2003); Patriarca, Chakraborti, and Kaski (2004); Chatterjee, Chakrabarti, and Manna (2004); Chatterjee, Chakrabarti, and Stinchcombe (2005); Chatterjee, Chakraborti, and Stinchcombe (2005); Patriarca, Chakraborti, Kimmo, and Germano (2005). The following substitution converts the Inequality Process into the Saved Wealth Model (apart from the random ω factor in Chatterjee et al, 2004 and subsequent papers):

$$d_{it} \rightarrow \epsilon_{it}$$

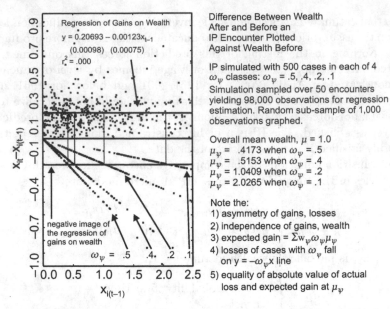

Fig. 6. The scattergram of wealth changes in the population of particles from time $t-1$ to t plotted against wealth at time $t-1$.

where ϵ_{it} is a continuous, uniform i.i.d random variate with support at $[0.0, 1.0]$.

2.2 The Macro-Model of the Inequality Process (IP)

The macro model of the Inequality Process (IP) is a gamma probability density function (pdf), $f_{\psi t}(x)$, a model of the wage income, x, of workers at the same level of education, the ψth at time t. The macro model approximates the stationary distribution of wealth of the IP's micro model. The macro model was developed in a chain of papers (Angle, 1993, 1996–2001, 2002b-2006). The IP's macro model in the ω_ψ equivalence class is:

$$f_{\psi t}(x) = \frac{\lambda_{\psi t}^{\alpha_\psi}}{\Gamma(\alpha_\psi)} x^{\alpha_\psi - 1} \exp(-\lambda_{\psi t} x) \tag{2}$$

or in terms of the IP's parameter in the ω_ψ equivalence class:

$$f_{\psi t}(x) = \exp\left[\left(\frac{1-\omega_\psi}{\omega_\psi}\right)\ln\left(\frac{1-\omega_i}{\tilde{\omega}_t \mu_t}\right)\right]$$
$$\times \exp\left[-\ln\Gamma\left(\frac{1-\omega_\psi}{\omega_\psi}\right) + \left(\frac{1-2\omega_i}{\omega_i}\right)\ln(x) - \left(\frac{1-\omega_i}{\tilde{\omega}_t \mu_t}\right)x\right] \tag{3}$$

where:

$\alpha_\psi \equiv$ the shape parameter of the gamma pdf that approximates the distribution of wealth, x, in the ω_ψ equivalence class, intended to model the wage income distribution of workers at the ψth level of education regardless of time; $\alpha_\psi > 0$

$$\alpha_\psi \approx \frac{1 - \omega_\psi}{\omega_\psi} \qquad (4)$$

and:

$\lambda_{\psi t} \equiv$ scale parameter of distribution of the gamma pdf that approximates the distribution of wealth, x, in the ω_ψ equivalence class, intended to model the wage income distribution of workers at level ψ of education in a labor force with a given unconditional mean of wage income and a given harmonic mean of ω_ψ's at time t;

$\lambda_{\psi t} > 0$

$$\lambda_{\psi t} \approx \frac{(1 - \omega_\psi)\left(\frac{u_{1t}}{\omega_1} + \ldots + \frac{u_{\psi t}}{\omega_\psi} + \ldots + \frac{u_{\Psi t}}{\omega_\Psi}\right)}{\mu_t} \approx \frac{(1 - \omega_\psi)}{\tilde{\omega}_t \mu_t} \qquad (5)$$

where:

μ_t = unconditional mean of wage income at time t

$\tilde{\omega}_t$ = harmonic mean of the ω_ψ's at time t.

and μ_t and the $u_{\psi t}$'s are exogenous and the sole source of change in a population of particles where Ψ ω equivalence classes are distinguished. Consequently, the dynamics of (2), the IP's macro model, are exogenous, that is, driven by the product $(\tilde{\omega}_t \mu_t)$ and expressed as a scale transformation, i.e., via $\lambda_{\psi t}$. Figure 7 shows the shapes of a gamma probability density function (pdf) for a fixed scale parameter and several values of the shape parameter, α_ψ. Figure 7 shows that if the IP's meta-theory is correct, more education, operationalized as smaller ω_ψ earns a worker a place in a wage income distribution with a larger α_ψ, a more centralized distribution, whose mean, equal to $\alpha_\psi / \lambda_{\psi t}$, is larger than that of the worker with less education.

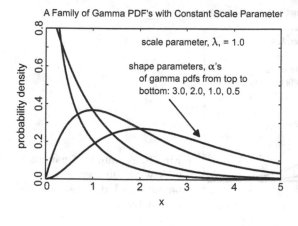

A Family of Gamma PDF's with Constant Scale Parameter

scale parameter, λ, = 1.0

shape parameters, α's of gamma pdfs from top to bottom: 3.0, 2.0, 1.0, 0.5

probability density

x

Fig. 7. A family of gamma pdfs with different shape parameters but the same scale parameter, 1.0.

Comparison of Figs. 7, 8, and 9 show the consequences of change in the gamma scale parameter on a gamma distribution holding the shape parameters constant. A decrease in λ_ψ stretches the mass of the pdf to the right over larger x's as in Fig. 8, increasing all percentiles of x. Compare Fig. 8 to Fig. 7. In the IP, particles circulate randomly within the distribution of wealth of their ω_ψ equivalence class, so an increase in μ_t of a magnitude sufficient to increase the product $(\tilde{\omega}_t \mu_t)$ and decrease $\lambda_{\psi t}$ may not mean that each and every particle in the ω_ψ equivalence class increases its wealth, although all the percentiles increase. $\tilde{\omega}_t$ is expected to decrease given a rising level of education in the U.S. labor force. μ_t, the unconditional mean of wage income, rose irregularly in the U.S. in the last four decades of the 20th century.

If proportional increase in μ_t offsets proportional decrease in $\tilde{\omega}_t$ then the product $(\tilde{\omega}_t \mu_t)$ increases, $\lambda_{\psi t}$ decreases, and the IP's macro model implies that wage income distribution is stretched to the right as in Fig. 8 with all percentiles of wage income increasing. However, if the product $(\tilde{\omega}_t \mu_t)$ decreases, then $\lambda_{\psi t}$ increases and the IP's macro model predicts that the wage income distribution is compressed to the left, that is, its mass is moved over smaller

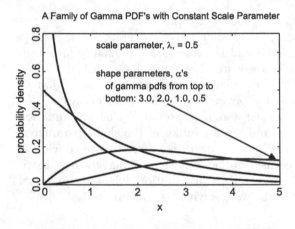

Fig. 8. A family of gamma pdfs with different shape parameters but the same scale parameter, 0.5.

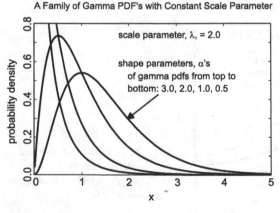

Fig. 9. A family of gamma pdfs with different shape parameters but the same scale parameter, 2.0.

wage income amounts and its percentiles decrease as in Fig. 9 by comparison to Figs. 7 and 8.

The product $(\bar{\omega}_t \mu_t)$ is estimated in the fit of the IP's macro model to the 43 distributions of wage income conditioned on education in the U.S. according to the March CPS' of 1962 through 2004, which collected data on wage incomes in 1961 through 2003. Six levels of worker education level have been distinguished. See Table 1. There are 43 X 6 = 258 partial distributions to be fitted. Also fitted are 258 median wage incomes, one for each partial distribution fitted. See Appendix B. Each partial distribution has fifteen relative frequency bins, each $10,000 (in constant 2003 dollars) wide, e.g., $1–$10,000, $10,001–$20,000, etc, for a total of 258 X 15 = 3,870 x (income), y (relative frequency) pairs to be fitted by the IP's macro model which has six degrees of freedom, the six values of ω_ψ estimated. The fits are simultaneous. The fitting criterion is the minimization of weighted squared error, i.e., nonlinear least squares. The weight on each partial distribution in each year in the fit is $u_{\psi t}$, the proportion of the labor force with ψth level of education in that year. A search is conducted over the parameter vector via a stochastic search algorithm that is a variant of simulated annealing to find the six values that minimize squared error. The squared correlation between the 3,870 observed and expected relative frequencies is .917. Table 1 displays the estimated parameters and their bootstrapped standard errors. Note that the estimated ω_ψ's scale inversely with level of education as predicted by the IP's meta-theory. Figure 10 displays the IP macro model's fit to the six partial distributions of wage income by level of education in 1981.

Separately, 258 gamma pdfs, each with two unconstrained parameters, were also fitted to each of the 258 partial distributions, a 516 parameter fit. These fits were done to create an alternative model to baseline how much less well the IP's macro model did than unconstrained gamma pdf fits to the same data set. The squared correlation between the 3,870 observed and expected relative frequencies under this alternative model is .957. Thus the IP's macro model fits the data almost as well as the unconstrained gamma pdf alternative

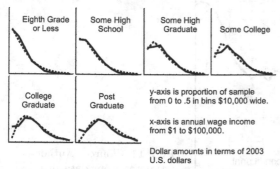

Distribution of U.S. Annual Wage Income
Conditioned on Education in 1981: Solid Curves
Fitted Macro Model: Dashed Curves

Fig. 10. Source: author's estimates based on March, CPS data.

Table 1. Estimates of the Parameters of the IP's Macro-Model

Highest Level of Education	ω_ψ estimated by fitting the macro-model to 258 partial distributions (43 years X 6 levels of education)	bootstrapped standard error of ω_ψ (100 re-samples)	estimate of α_ψ corresponding to ω_ψ
eighth grade or less	0.4524	.0009582	1.1776
some high school	0.4030	.0006159	1.4544
high school graduate	0.3573	.0004075	1.7924
some college	0.3256	.0005033	2.0619
college graduate	0.2542	.0007031	2.7951
post graduate education	0.2084	.0005216	3.6318

model although the IP's macro model uses only 6 degrees of freedom and the alternative model 516.

Figure 11 shows that the unconditional mean of wage income in the U.S. increased substantially in the 1960's and again in the 1990's in constant 2003 dollars. There was a smaller move upward in the early to mid 1980's. However between the early 1970's and mid-1990's there were small declines and small increases that netted each other out, i.e., the unconditional mean of wage income in the U.S. did not increase in constant dollar terms for over two decades. The IP's macro model implies that the scale factor of wage income at each level of education in the labor force is driven by the product $(\tilde{\omega}_t \mu_t)$. In the model bigger $(\tilde{\omega}_t \mu_t)$ stretches the distribution of wage incomes at each level of education to the right over larger wage incomes. μ_t has to increase

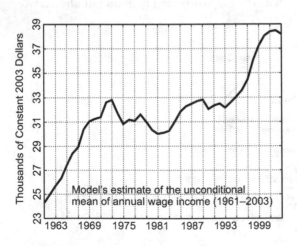

Fig. 11. Source: Author's estimates from data of the March Current Population Survey.

proportionally more than $\tilde{\omega}_t$ decreases for all percentiles of the distribution conditioned on education to increase.

$\tilde{\omega}_t$ is the harmonic mean of the ω_ψ's at each time point. The proportion each ω_ψ equivalence class forms of the population, $u_{\psi t}$, changes as the proportion of workers at a given educational level changes in the labor force. From 1961–2003 the level of education of the U.S. labor force rose substantially. See Fig. 12. Given the ω_ψ's estimates in Table 1, $\tilde{\omega}_t$ decreases 1961–2003 as the level of education rises in the U.S. labor force. Figure 13 displays the course of $\tilde{\omega}_t$ from 1961 through 2003, a steady decline throughout.

Figure 14 graphs the estimated product $(\tilde{\omega}_t \mu_t)$ over time. Note that decline between the early 1970's and mid-1990's was much larger proportionally than in Fig. 11, the time-series of μ_t. That means that the U.S. wage income distribution conditioned on education was compressed substantially to the left over smaller wage incomes from 1976 through 1983 in a much more pronounced way than Fig. 11 implies. Figure 11 incorporates the positive effect on the unconditional mean of the rise in education level in the U.S. labor

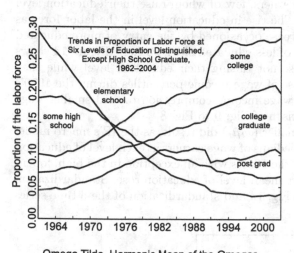

Fig. 12. Source: Author's estimates from data of the March Current Population Survey.

Omega Tilde, Harmonic Mean of the Omegas

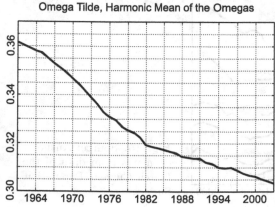

Fig. 13. Source: Author's estimates from data of the March Current Population Survey.

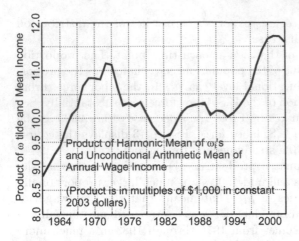

Fig. 14. Product of the unconditional mean, μ_t, and the harmonic mean of the ω_ψ's, $\tilde{\omega}_t$. Source: Author's estimates from data of the March Current Population Survey.

force from 1961 through 2003. Figure 14 shows the negative effect of the rise in educational level on wage earners, few of whom raise their education level while they work for a living. The rise in education level in the labor force as a whole is due to the net increase occasioned by the entry of more educated younger workers and the exit of less educated older workers. Figure 14 shows that most wage earners, those not raising their education level while they worked, experienced a decrease of wage income percentiles during the 1970's and early 1980's, the sort of wage income compression toward smaller wage incomes shown in the comparison of Fig. 9 to Fig. 8.

Figure 15 confirms that smaller $(\tilde{\omega}_t \mu_t)$ did result, as the IP's macro model implies, in downturns in the medians of wage earners at each level of education from the early 1970's through the 1990's with an exception in the high, 'open-end' category of education. Its mean level of education rose. Standardization of the time-series of $(\tilde{\omega}_t \mu_t)$ in Fig. 14 and standardization of the 6 time-series

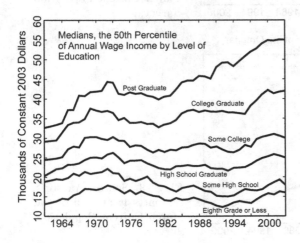

Fig. 15. Source: Author's estimates from data of the March Current Population Survey.

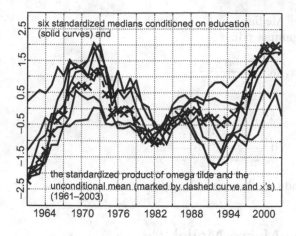

Fig. 16. Source: Author's estimates from data of the March Current Population Survey.

of conditional median wage incomes in Fig. 15 allow direct observation of how closely these six are associated with $(\tilde{\omega}_t \mu_t)$. Figure 16 shows the graphs of the 7 standardized time-series. The time-series of the standardized $(\tilde{\omega}_t \mu_t)$'s is marked by X's. The 6 standardized medians track the standardized $(\tilde{\omega}_t \mu_t)$'s. Table 2 shows that 4 of the 6 time-series of conditional medians are more closely correlated with the product $(\tilde{\omega}_t \mu_t)$ than with the unconditional mean, μ_t, alone. The IP's macro model implies a larger correlation between the time-series of median wage income conditioned on education and $(\tilde{\omega}_t \mu_t)$ than with μ_t alone. This inference follows from Doodson's approximation formula for the median of the gamma pdf, $f_{\psi t}(x)$, $x_{(50)\psi t}$ (Weatherburn, 1947:15 [cited in Salem and Mount, 1974: 1116]):

Table 2. Estimated Correlations Between Time-Series

Highest Level of Education	correlation between $\tilde{\omega}_t \mu_t$ and median wage income at a given level of education	correlation between the unconditional mean μ_t and median wage income at a given level of education
eighth grade or less	.5523	.1837
some high school	.1573	−.2874
high school graduate	.8729	.5885
some college	.9279	.7356
college graduate	.9042	.9556
post graduate education	.7776	.9575

$$\text{Mean} - \text{Mode} \approx 3 \,(\text{Mean} - \text{Median})$$

in terms of a two parameter gamma pdf:

$$x_{(50)\psi t} \approx \frac{3\alpha_\psi - 1}{3\lambda_{\psi t}}$$

and given (4) and (5):

$$x_{(50)\psi t} \approx \left(\frac{1 - \frac{4}{3}\omega_\psi}{1 - \omega_\psi}\right)\left(\frac{\tilde{\omega}_t \mu_t}{\omega_\psi}\right) \tag{6}$$

a constant function of the conditional mean, $(\tilde{\omega}_t \mu_t)/\omega_\psi$.

3 The Dynamics of the Macro Model of the Inequality Process (IP)

The dynamics of the Inequality Process (IP)'s macro-model of the wage income distribution of workers at the same level of education are driven exogenously by change in $(\tilde{\omega}_t \mu_t)$:

$$\frac{\partial f_{\psi t}(x)}{\partial (\tilde{\omega}_t \mu_t)} = f_{\psi t}(x) \; \lambda_{\psi t} \left(\frac{x - \mu_{\psi t}}{\tilde{\omega}_t \mu_t}\right)$$

$$= f_{\psi t}(x) \; \frac{(1 - \omega_\psi)}{(\tilde{\omega}_t \mu_t)^2} \, (x - \mu_{\psi t}) \tag{7}$$

where, the conditional mean of wealth in the ω_ψ equivalence class, μ_t, is:

$$\mu_{\psi t} = \frac{\alpha_\psi}{\lambda_{\psi t}} \approx \frac{\tilde{\omega}_t \mu_t}{\omega_\psi} \tag{8}$$

In (7), as $(\tilde{\omega}_t \mu_t)$ increases, $f_{\psi t}(x_0)$ decreases to the left of the conditional mean, $\mu_{\psi t}$, i.e., for $x_0 < \mu_{\psi t}$. $f_{\psi t}(x_0)$ increases to the right of the conditional mean $\mu_{\psi t}$, i.e., for $x_0 > \mu_{\psi t}$. So an increase in $(\tilde{\omega}_t \mu_t)$ simultaneously thins the left tail of the distribution of x, wealth, in the ω_ψ equivalence class and thickens the right tail. (7) implies that the probability mass in the left and right tails, defined as the probability mass over $x_0 < \mu_{\psi t}$ and $x_0 > \mu_{\psi t}$ respectively, must vary inversely if $(\tilde{\omega}_t \mu_t)$ changes. Thus, the macro-model of the Inequality Process (IP) squarely contradicts the hypothesis that a wage distribution conditioned on education can become U-shaped via a simultaneous thickening of the left and right tails, what the literature on the 'hollowing out' of the U.S. distribution of wage incomes asserts.

Given that in the IP's macro-model all change is exogenous, due to $(\tilde{\omega}_t \mu_t)$, the forward difference, $f_{\psi t}(x_0) - f_{\psi(t-1)}(x_0)$, at a given x_0 can be approximated via Newton's approximation as:

$$f_{\psi t}(x_0) - f_{\psi(t-1)}(x_0) \approx f_{\psi(t-1)}(x_0) + f'_{\psi(t-1)}(x_0) \left((\tilde{\omega}_t \mu_t) - (\tilde{\omega}_{t-1} \mu_{t-1}) \right)$$
$$- f_{\psi(t-1)}(x_0)$$

$$\approx f_{\psi(t-1)}(x_0) \cdot \frac{(1 - \omega_\psi)}{(\tilde{\omega}_{t-1} \mu_{t-1})} \cdot \left(x_0 - \mu_{\psi(t-1)} \right)$$
$$\times \left(\frac{\tilde{\omega}_t \mu_t}{\tilde{\omega}_{t-1} \mu_{t-1}} - 1 \right)$$

$$\approx f_{\psi(t-1)}(x_0) \cdot \lambda_{\psi(t-1)} \cdot \left(x_0 - \mu_{\psi(t-1)} \right)$$
$$\times \left(\frac{\tilde{\omega}_t \mu_t}{\tilde{\omega}_{t-1} \mu_{t-1}} - 1 \right) \tag{9}$$

(9) says that the forward difference, $f_{\psi t}(x_0) - f_{\psi(t-1)}(x_0)$, of relative frequencies of the same x_0 is proportional to $f_{\psi(t-1)}(x_0)$, the scale parameter at time $t - 1$, $\lambda_{\psi(t-1)}$, the signed difference between x_0 and the conditional mean, $\mu_{\psi(t-1)}$ at time $t - 1$, and the signed proportional increase (positive) or proportional decrease (negative) in the product $(\tilde{\omega}_{t-1} \mu_{t-1})$. (9) implies little change in the relative frequency of wage income in the vicinity of the conditional mean. It also implies that the $(x_0 - \mu_{\psi(t-1)})$ term can become largest in absolute value in the extreme right tail, i.e., for the largest x_0, since the absolute value of the difference $(x_0 - \mu_{\psi(t-1)})$ is greater for the maximum x_0, typically more than three times the mean, than it is for the minimum x_0, which is very nearly one mean away from the mean. However, the forward difference, $(f_{\psi t}(x_0) - f_{\psi(t-1)}(x_0))$, will still be forced down toward zero in the far right tail when $(\tilde{\omega}_{t-1} \mu_{t-1})$ increases because the RHS of (9) is multiplied by $(f_{\psi(t-1)}(x_0))$ which becomes small quickly as x_0 becomes large. So the forward difference becomes small in the far right tail even when $(\tilde{\omega}_{t-1} \mu_{t-1})$ increases despite the fact that $(x_0 - \mu_{\psi(t-1)})$ reaches its positive maximum for the maximum x_0. However, Newton's approximation to the ratio, $(f_{\psi t}(x_0)/f_{\psi(t-1)}(x_0))$, reflects the full effect of $(x_0 - \mu_{\psi(t-1)})$ on growth in the density of the far right tail when $(\tilde{\omega}_{t-1} \mu_{t-1})$ increases.

Given that in the IP's macro-model change is exogenous, due to $(\tilde{\omega}_t \mu_t)$, the ratio, $f_{\psi t}(x_0)/f_{\psi(t-1)}(x_0)$, is approximated via Newton's approximation as:

$$\frac{f_{\psi t}(x_0)}{f_{\psi(t-1)}(x_0)} \approx \frac{f_{\psi(t-1)}(x_0) + f'_{\psi(t-1)}(x_0) \left((\tilde{\omega}_t \mu_t) - (\tilde{\omega}_{t-1} \mu_{t-1}) \right)}{f_{\psi(t-1)}(x_0)}$$

$$\approx \left[1 + \left[(x_0 - \mu_{\psi(t-1)}) \left(\frac{1 - \omega_\psi}{\tilde{\omega}_{t-1} \mu_{t-1}} \right) \left(\frac{\tilde{\omega}_t \mu_t}{\tilde{\omega}_{t-1} \mu_{t-1}} - 1 \right) \right] \right] \tag{10}$$

The bigger the $(x_0 - \mu_{\psi(t-1)})$ term is in the right tail, the greater is the ratio $f_{\psi t}(x_0)/f_{\psi(t-1)}(x_0)$ when $(\tilde{\omega}_t \mu_t)$ increases. Figure 1, showing the surge in wage income nouveaux riches, graphs the empirical analogue of the ratio $f_{\psi t}(x_0)/f_{\psi(t-1)}(x_0)$. (10) is descriptive of Fig. 1. Note that according to (10), in the right tail where $x_0 > \mu_{\psi(t-1)}$, the difference $(x_0 - \mu_{\psi(t-1)})$ for x_0 fixed becomes smaller as the conditional mean, $\mu_{\psi(t-1)}$, increases with increasing $(\tilde{\omega}_t \mu_t)$, implying a deceleration in the rate of increase of the ratio $f_{\psi t}(x_0)/f_{\psi(t-1)}(x_0)$ for a given increase in $(\tilde{\omega}_t \mu_t)$. This deceleration is evident in Fig. 1.

The expression for forward proportional change is that of the RHS of (10) minus 1.0:

$$
\begin{aligned}
\frac{f_{\psi t}(x_0) - f_{\psi(t-1)}(x_0)}{f_{\psi(t-1)}(x_0)} &\approx \frac{f'_{\psi(t-1)}(x_0)\left((\tilde{\omega}_t \mu_t) - (\tilde{\omega}_{t-1}\mu_{t-1})\right)}{f_{\psi(t-1)}(x_0)} \\
&\approx \left[(x_0 - \mu_{\psi(t-1)})\left(\frac{1 - \omega_\psi}{\tilde{\omega}_{t-1}\mu_{t-1}}\right)\left(\frac{\tilde{\omega}_t \mu_t}{\tilde{\omega}_{t-1}\mu_{t-1}} - 1\right)\right] \\
&\approx \lambda_{\psi(t-1)}\left(x_0 - \mu_{\psi(t-1)}\right)\left(\frac{\tilde{\omega}_t \mu_t}{\tilde{\omega}_{t-1}\mu_{t-1}} - 1\right) \qquad (11)
\end{aligned}
$$

and it has like (9) the property that it changes sign according to whether x_0 is greater than or less than the conditional mean, $\mu_{\psi t}$, and whether $\tilde{\omega}_t \mu_t$ has increased or decreased. For example, in the right tail of the distribution, i.e., $x_0 > \mu_{\psi t}$, when $\tilde{\omega}_t \mu_t$ increases, forward proportional change in the distribution, $f_{\psi t}(x_0)$, is positive. Forward proportional change in the distribution is a product of the three factors on the RHS of (11). Forward proportional change in the distribution, $f_{\psi t}(x_0)$, in (11) is a linear function of the difference $(x_0 - \mu_{\psi(t-1)})$ and can, since maximum x_0 can be at least three times as far from the mean as minimum x_0, forward proportional growth in the extreme right of the right tail when $\tilde{\omega}_t \mu_t$ increases is greater than at any other income amount. In other words, the IP's macro model implies rapid growth in the population of wage income _nouveaux riches_ whenever $(\tilde{\omega}_t \mu_t)$ increases. One would expect that purveyors of goods and services priced for people with large wage incomes might see their market experiencing explosive growth whenever the product $(\tilde{\omega}_t \mu_t)$ increases.

3.1 The Implied Dynamics of the IP's Macro Model for the Unconditional Distribution of Wage Income

The IP's macro model of the unconditional wage income distribution, a mixture of gamma pdf's, $f_t(x_0)$, is:

$$
f_t(x_0) = u_{1t}f_{1t}(x_0) + \ldots + u_{\psi t}f_{\psi t}(x_0) + \ldots + u_{\Psi t}f_{\Psi t}(x_0) \qquad (12)
$$

where:
$f_{\psi t}(x_0) \equiv$ IP's macro model of distribution of wealth in the ω_ψ equivalence class at time t;
$u_{\psi t} \equiv$ proportion of particles in the ω_ψ equivalence class at time t, the mixing weights.

The dynamics of (12), the unconditional relative frequency of wage income, are driven by $(\tilde{\omega}_t \mu_t)$ as in (7) and also by the direct effect of the $u_{\psi t}$'s:

$$\frac{\partial f_t(x_0)}{\partial(\tilde{\omega}_t\mu_t)} = \sum_\psi \left(u_{\psi t} \, f_{\psi t}(x_0) \, \frac{(1 - \omega_\psi)}{(\tilde{\omega}_t\mu_t)^2} \, (x_0 - \mu_{\psi t}) \right) \tag{13}$$

(12) is a gamma pdf mixture; a gamma mixture is not, in general, a gamma pdf. While (12) shares many properties of (2) in the ω_ψ equivalence class, it has others as well, namely the direct effect of change in the proportions, the $u_{\psi t}$'s, in each ω_ψ equivalence class. Figure 12 shows that the $u_{\psi t}$'s of larger ω_ψ's (those of the less well educated, e.g. workers without a high school diploma) decreased between 1962 and 2004 while the $u_{\psi t}$'s of the smaller ω_ψ's (those of the more educated, e.g., with at least some post-secondary school education) increased. This change in the $u_{\psi t}$'s implies that $\tilde{\omega}_t$ decreased in this period, as Fig. 13 shows.

The implications for the right tail of the conditional distribution, $f_{\psi t}(x_0)$, in (9), (10), and (11), as the product $(\tilde{\omega}_t\mu_t)$ increases, carry through for the dynamics of the right tail of the unconditional distribution, $f_t(x_0)$, for $x_0 >$ $\mu_{\phi t}$ where $\mu_{\phi t}$ is the mean of x in the ω_ϕ equivalence class where ω_ϕ is the

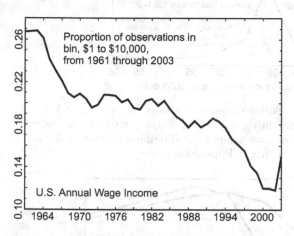

Fig. 17. Relative frequencies of incomes \$1–\$10,000 in the unconditional distribution. Source: Author's estimates from data of the March Current Population Survey.

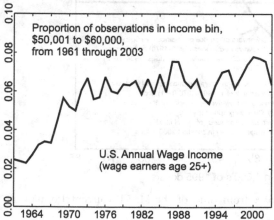

Fig. 18. Relative frequencies of incomes \$50,001–\$60,000 in the unconditional distribution. Source: Author's estimates from data of the March Current Population Survey.

204   J. Angle

minimum ω, (and consequently $\mu_{\phi t}$ is the maximum mean of any ω equivalence class), and for the dynamics of the left tail of the unconditional distribution, $f_t(x_0)$, for $x_0 < \mu_{\theta t}$ where $\mu_{\theta t}$ is the mean of x in the ω_θ equivalence class, where ω_θ is the maximum ω in the population (and consequently $\mu_{\theta t}$ is the minimum mean of any ω equivalence class). Thus, as $(\tilde{\omega}_t \mu_t)$ and the u_ψ in equivalence classes with smaller ω_ψ's increase, (13) implies that the left tail thins and the right tail thickens. Figures 17 and 18 show that such is the case in the left tail bin, \$1–\$10,000, and the right tail bin, \$50,001–\$60,000 (both in

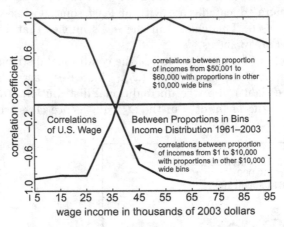

Fig. 19. Correlations between relative frequency in wage income bin in left tail, \$1–\$10,000, and relative frequency in right tail bin, \$50,001–\$60,000, and relative frequencies in all the other income bins around the distribution. Source: Author's estimates from data of the March Current Population Survey.

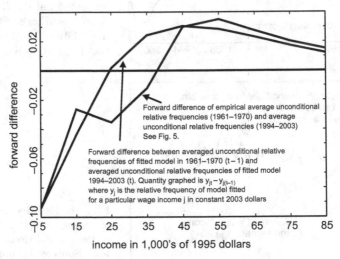

Fig. 20. Source: Author's estimates from data of the March Current Population Survey.

constant 2003 dollars). Figure 19 shows how each relative frequency (that in the bin $1–$10,000 and that in the bin, $50,001–$60,000) has a large positive correlation with other relative frequencies in the same tail and a large negative correlation with relative frequencies in the other tail. For example, the relative frequencies in the bins $1–$10,000 and $50,001–$60,000 have a nearly a perfect negative correlation with each other. Both relative frequencies, as one would expect given (13), have a near zero correlation with relative frequencies close to the unconditional mean of wage income.

Figure 20 shows that the unconditional forward difference of wage incomes between the average of the relative frequencies in the period 1961 to 1970 and the average of the relative frequencies in the period 1994–2003 largely overlaps the fitted forward difference between the expected relative frequencies in these two periods at the beginning and end of the time series. The time averaging is done to smooth out the pronounced frequency spiking in these data. See Angle (1994) for a discussion of frequency spiking in the wage income observations collected by the March Current Population Survey.

4 Conclusions

The IP's macro model fits the distribution of U.S. wage income conditioned on education 1961–2003. It also accounts for one of the quirkier time-series of scalar statistics of U.S. wage income in the same time period: the more rapid growth in the relative frequency of the larger wage income in the right tail of the distribution, that is, among wage incomes greater than mean wage income. Figure 20 shows that the IP's macro model accounts for how the relative frequencies of wage income changed between 1961–1970 and 1994–2003. Figure 21 shows why, in particular, for large wage incomes (defined in constant 2003 dollars): the expected frequencies of large wage incomes under the IP's macro model track the observed frequencies of large wage incomes closely.

The observed relative frequencies are estimated from reports of personal annual wage income in the micro-data file, the individual person records, of the March Current Population Survey (CPS), in 'public use' form, i.e., with personal identifiers stripped from the file. The March CPS is a survey of a large sample of households in the U.S. conducted by the U.S. Bureau of the Census. In the March CPS, a respondent answers questions posed by a Census Bureau interviewer about members of the household. There is a question about the annual wage income of each member of the household in the previous year. See Fig. 4 for estimates of the distribution of annual wage income 1961–2003. All dollar amounts have been converted to constant 2003 dollars.

The U.S. Census Bureau has evaluated the adequacy of its wage income question in the March Current Population Survey (CPS) and acknowledged that respondents, on average, underestimate the wage income they report (Roemer, 2000: 1). Roemer writes "Many people are reluctant to reveal their incomes to survey researchers and this reluctance makes such surveys par-

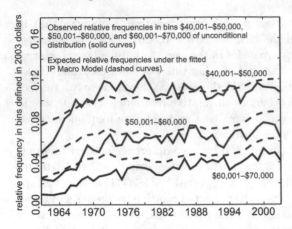

Fig. 21. Unconditional right tail relative frequencies 1961–2003 (*solid curves*) and estimated right tail relative frequencies under the fitted IP Macro Model (*dashed curves*). Source: Author's estimates from data of the March Current Population Survey.

ticularly prone to response errors." Roemer (2000: 17–21) reports that these underestimates are least biased downward for wage incomes near the median but seriously biased downward for large wage incomes. So it is not a problem for the IP's macro model if it overestimates the relative frequency of large wage incomes slightly, particularly very large wage incomes, as you can see it does in Fig. 21.

The macro model of the Inequality Process is a gamma probability density function (pdf) whose parameters are derived from the micro model of the Inequality Process and expressed in terms of its parameters. See (2) through (5). The dynamics of this model are expressed in terms of the gamma scale parameter, (5), of this model. (5) says that the model is driven exogenously by the product $(\tilde{\omega}_t \mu_t)$ through its scale parameter, $\lambda_{\psi t}$. $(\tilde{\omega}_t \mu_t)$ is a function of the distribution of education in the labor force at time t and the unconditional mean of wage income, μ_t, at time t. $\tilde{\omega}_t$ is the harmonic mean of the estimated IP parameters, the ω_ψ's. These are estimated in the fitting of the IP's macro model to the distribution of wage income conditioned on education, 1961–2003. The ω_ψ's also enter the formula by which the unconditional mean, μ_t, is estimated from sample conditional medians under the hypothesis that wage income is gamma distributed. The u_ψ's, the proportions in each ω_ψ equivalence class, by hypothesis the fraction of the labor force at a particular level of education, also enter the formula by which μ_t is estimated from sample conditional medians. The IP's macro model fits the distribution of wage income in the U.S., 1961–2003, well.

4.1 The Dynamics of the Wage Income Distribution When $(\tilde{\omega}_t \mu_t)$ Increases: A Stretching, Not a 'Hollowing Out'

Not a 'Hollowing Out'

When $(\tilde{\omega}_t \mu_t)$ increases, the distribution of wage income stretches to the right over larger wage incomes, as in the comparison of Fig. 8 to Fig. 7. Figure 8 is the graph of gamma pdf's with different shapes but the same scale parameter. Figure 8 has gamma pdfs with the same shape parameters but a different scale parameter, one that is half that of Fig. 7. The gamma pdf's of Fig. 8 look stretched to the right. When $(\tilde{\omega}_t \mu_t)$ decreases, the wage income distribution is compressed to the left over smaller wage incomes, as in the comparison of Fig. 9 to Fig. 7. These effects are deduced from the IP's macro model in (9). The last term in the product on the RHS of (9) is positive when $(\tilde{\omega}_t \mu_t) > (\tilde{\omega}_{t-1} \mu_{t-1})$, negative when $(\tilde{\omega}_t \mu_t) < (\tilde{\omega}_{t-1} \mu_{t-1})$, meaning that when $(\tilde{\omega}_t \mu_t)$ increases, the right tail thickens, the left tail thins, and vice versa when $(\tilde{\omega}_t \mu_t)$ decreases. While the IP's micro model is time-reversal asymmetric, its macro model is time-reversal symmetric. The IP's macro model implies in (10), and (11) that growth in the relative frequency of large wage incomes, i.e., the thickness of the right tail of the wage income distribution, is greater, the larger the wage income, i.e., the farther to the right in the tail, when $(\tilde{\omega}_t \mu_t)$ increases.

So the IP's macro model accounts for the surge in the far right tail of the wage income distribution in the U.S., the appearance of wage income nouveaux riches, as $(\tilde{\omega}_t \mu_t)$ increased from 1961 through 2003. See Figs. 1, 5, 20, and 21. The IP's macro model implies that the right tail of the wage income distribution thickened as $(\tilde{\omega}_t \mu_t)$ increased from 1961 through 2003 and the left tail of the distribution thinned. The empirical evidence bears out this implication of the IP's macro model, but contradicts the interpretation in the labor economics literature that the thickening of the right tail of the wage income distribution represented a 'hollowing out' of the wage income distribution, that is, a simultaneous thickening in the left and right tails of the distribution at the expense of the relative frequency of wage incomes near the median of the distribution, as illustrated conceptually in Fig. 3.

As you can see in Fig. 4, the unconditional distribution of wage income thinned in its left tail and thickened in its right from 1961 through 2003. Figure 17 shows how the relative frequency of wage incomes from $1–$10,000 (constant 2003 dollars) decreased from 1961 through 2003, although not monotonically, while Fig. 18 shows how the relative frequency of wage incomes from $50,001–$60,000 (constant 2003 dollars) increased from 1961 through 2003, although not monotonically. $50,001 in 2003 dollars is greater than the unconditional mean of wage income from 1961 through 2003, so the wage income bin $50,001–$60,000 was in the right tail the entire time. If there is any remaining question of what was happening elsewhere in the distribution, it is answered by Fig. 19 which shows the correlation between the relative frequency in income bin $1–$10,000 with that of every other income bin. The relative frequency of this extreme left tail bin was positively correlated with the relative frequency in the other left tail bin, had almost no correlation with relative frequency of mean wage income, and a large negative correlation with relative frequencies of all the right tail income bins. Figure 19 also shows the correlations of the relative frequency of the income bin $50,001–$60,000 with relative frequencies in other bins around the distribution. These correlations are a near mirror image of the correlations of the left tail bin \$1–\$10,000. The relative frequency of income bin \$50,001–\$60,000 has a high positive correlation with the relative frequencies of other right tail income bins, near zero correlation with the relative frequency of mean income, and a large negative correlation with the relative frequencies of left tail wage income bins. Figure 20 shows that the relative frequency of wage incomes smaller than the mean decreased between 1961 and 2003 while those greater than the mean increased. There is no doubt that the relative frequencies of the left tail of the wage income distribution vary inversely with the relative frequencies of the right tail, just as the IP's macro model implies, in contradiction of the 'hollowing out' hypothesis.

A Stretching of the Distribution When $(\tilde{\omega}_t \mu_t)$ Increases

This paper has focused on how the relative frequency of a wage income of a given size changes when $(\tilde{\omega}_t \mu_t)$ increases because it is algebraically transparent. The algebra indicates more rapid growth in the relative frequency of the larger wage income in the right tail of the distribution. However, a clearer demonstration of how the IP's macro model and the empirical wage income change when $(\tilde{\omega}_t \mu_t)$ increases is in the dynamics of the percentiles of wage income, that is, not how the relative frequency of a particular fixed wage income in constant dollars, x_0, changes, but rather how the percentiles of the distribution change. Figure 2 shows that the 90th percentile of wage income increased more in absolute terms than the 10th percentile between 1961 and 2003, i.e., the distribution stretched farther to the right over larger wage incomes in its right tail than its left. Does the same occur with the 10th and 90th percentiles of the IP's macro model of the unconditional distribution of wage income? This demonstration requires numerical integration and so is less transparent algebraically than inspecting the algebra of the model for the dynamics of the relative frequency of large wage incomes.

Figure 22 displays how well the percentiles of the model track the observed percentiles of wage income. The tendency to slightly overestimate the 90th percentile is not a problem given Roemer's (2000) evaluation of the accuracy of reporting of wage income data in the March CPS. In Fig. 22 the graphs of the unconditional percentiles of the IP's macro model and of empirical wage income as $(\tilde{\omega}_t \mu_t)$ increases show both distributions stretching to the right: the bigger the percentile, the more it increases in absolute constant dollars, what one would expect from the multiplication of all wage income percentiles by the same multiplicative constant, usually greater than 1.0, in each year between 1961 and 2006.

A percentile, $x_{(i)\psi t}$, of the IP's macro model, $f_{\psi t}(x)$, is:

$$\frac{i}{100} = \int_0^{x_{(i)\psi t}} \frac{\lambda_{\psi t}^{\alpha_\psi}}{\Gamma(\alpha_\psi)} x^{\alpha_\psi - 1} \exp(-\lambda_{\psi t} x) \, dx$$

where i is integer and i is less than or equal to 100. Figure 15 graphs the conditional medians, the 50th percentiles, $x_{(50)\psi t}$'s, from 1961 through 2003. Figure 16 shows that, when standardized, i.e., when their mean is subtracted from them and this difference is divided by their standard deviation, the transformed conditional medians have a time-series close to that of the standardization of $(\tilde{\omega}_t \mu_t)$. (6), Doodson's approximation to the median of a gamma pdf in terms of the IP's parameters, shows why: the median is approximately a constant function of $(\tilde{\omega}_t \mu_t)$. $(\tilde{\omega}_t \mu_t)$ enters $f_{\psi t}(x)$ as a gamma scale parameter transformation, via $\lambda_{\psi t}$. A scale transformation affects all percentiles multiplicatively, as in the comparison of Fig. 8 to figure 7. The gamma pdfs of Fig. 8 have the same shape parameters as those of Fig. 7. The difference between the two sets of graphs is that those of Fig. 8 have scale parameters, $\lambda_{\psi t}$, that are one half those of Fig. 7. The gamma pdfs of Fig. 8 have been stretched to the right over larger x's from where they were in Fig. 7. The IP's macro model implies this stretching to the right over larger wage incomes when the product $(\tilde{\omega}_t \mu_t)$ increases, which Fig. 14 shows it did from 1961 through 2003, although not monotonically so. A larger $(\tilde{\omega}_t \mu_t)$ results in a smaller gamma scale parameter, $\lambda_{\psi t}$, given (5).

So, the Inequality Process' (IP) macro model explains both the surge in the relative frequency of large wage incomes and the greater absolute increase in the greater percentile of wage incomes in the U.S., 1961–2003 as $(\tilde{\omega}_t \mu_t)$ increased. Since the $\tilde{\omega}_t$ term decreases with rising levels of education in the U.S. labor force, the condition of $(\tilde{\omega}_t \mu_t)$ increasing means that the unconditional mean of wage income, μ_t, grew more proportionally 1961–2003 than $\tilde{\omega}_t$ decreased. Since all percentiles of wage income grew as $(\tilde{\omega}_t \mu_t)$ increased, the surge in wage income nouveaux riches in the U.S. 1961–2003 was simply a visible indicator of generally rising wage incomes, hardly the ominous event it was made out to be by some in the scholarly literature and the popular press.

Appendix A: The March Current Population Survey And Its Analysis

The distribution of annual wage and salary income is estimated with data from the March Current Population Surveys (CPS) (1962–2002), conducted by the U.S. Bureau of the Census. One of the money income questions asked on the March CPS is total wage and salary income received in the previous calendar year. See Weinberg, Nelson, Roemer, and Welniak (1999) for a description of the CPS and its history. The CPS has a substantial number of households in its nationwide sample. The March Current Population Survey (CPS) provides the data for official U.S. government estimates of inequality of wage income as well as most of the labor economics literature on inequality of wage income in the U.S.

The present paper examines the civilian population of the U.S. that is 25+ in age and earns at least \$1 (nominal) in annual wage income. The age restriction to 25+ is to allow the more educated to be compared to the less educated. It is a conventional restriction in studies of the relationship of education to wage income. The data of the March CPS of 1962 through 2004 were purchased from Unicon Research, inc. (Unicon Research, inc, 2004; Current Population Surveys, March 1962–2004), which provides the services of data cleaning, documentation of variable definitions and variable comparability over time, and data extraction software. Unicon Research, inc was not able to find a copy of the March 1963 CPS public use sample containing data on education. Consequently, the distribution of wage and salary income received in 1962 (from the March 1963 CPS) conditioned on education is interpolated from the 1961 and 1963 (from the 1962 and 1964 March CPS').

All dollar amounts in the March CPS are converted to constant 2003 dollars using the U.S. Bureau of Economic Analysis National Income and Product Account Table 2.4.4 Price indexes for personal consumption expenditure by type of product [index numbers, 2000 = 100] http://www.bea.gov/bea/dn/nipaweb/TableView.asp#Mid [Last revised on 8/4/05].

Appendix B: Estimation

Estimation of Relative Frequencies

All estimates are weighted estimates. The weight associated with the jth observation in the tth year, u_{jt}^*, is:

$$u_{jt}^* = \frac{u_{jt}}{\sum_{i=1}^{n_t} u_{it}} \, n_t$$

where,

u_{jt} = the raw weight provided by the Census Bureau for observation j

n_t = the sample size in year t.

Estimation of the μ_t, the Unconditional Mean, from Sample Conditional Medians, $x_{(50)\psi t}$'s

While an unconditional sample mean of wage incomes in the March CPS can be directly estimated from the data, it is known to be an underestimate of the population unconditional mean, μ_t. The sampling frame of the March CPS does not sample large wage incomes at a higher rate than smaller wage incomes. Consequently, given the right skew of the distribution wage income dollars will be missed in the form of very large individual wage incomes biasing the sample mean of wage income downward. Further, the Census Bureau itself has concluded that even when a household with one or more large wage incomes falls into the sample, those wage income reports have a serious downward bias (Roemer, 2000:17–21). The sample median of wage incomes is robust against these problems of estimation. It is as well measured as any sample statistic of annual wage income.

The unconditional mean of the IP's macro model, μ_t, is estimated in terms of the sample conditional medians, the $x_{(50)\psi t}$'s, (the median wage income at the ψth level of education) and the $u_{\psi t}$'s, (the proportion of the labor force at the ψth level of education) using Doodson's approximation formula for the median of a gamma pdf, (Weatherburn, 1947:15 [cited in Salem and Mount, 1974]) as instantiated for the IP's macro model in (6), since:

$$\mu_t = u_{1t}\mu_{1t} + u_{2t}\mu_{2t} + \ldots + u_{\psi t}\mu_{\psi t} + \ldots + u_{6t}\mu_{6t}.$$

References

1. Angle J (1983) The surplus theory of social stratification and the size distribution of Personal Wealth. Proc. Am. Stat. Assoc., Social Statistics Section. Pp. 395–400. Alexandria, VA: Am. Stat. Assoc.
2. Angle J (1986) The surplus theory of social stratification and the size distribution of Personal Wealth. Social Forces 65:293–326.
3. Angle J (1993) Deriving the size distribution of personal wealth from 'the rich get richer, the poor get poorer'. J. Math. Sociology 18:27–46.
4. Angle J (1994) Frequency spikes in income distributions. Proc. Am. Stat. Assoc., Business and Economic Statistics Section, pp. 265–270.
5. Angle J (1996) How the gamma law of income distribution appears invariant under aggregation. J. Math. Sociology. 21:325–358.
6. Angle J (2002) The statistical signature of pervasive competition on wages and salaries. J. Math. Sociology. 26:217–270.
7. Angle J (2002) Modeling the dynamics of the nonmetro distribution of wage and salary income as a function of its mean. Proc. Am. Stat. Assoc., Business and Economic Statistics Section. [CD-ROM], Alexandria, VA: Am. Stat. Assoc.
8. Angle J (2003) The dynamics of the distribution of wage and salary income in the nonmetropolitan U.S.. Estadistica 55:59–93.
9. Angle J (2005) U.S. wage income since 1961: the perceived inequality trend. Paper presented to the Population Assoc. of America meetings, March-April 2005, Philadelphia, Pennsylvania, USA. http://paa2005.princeton.edu/download.aspx?submissionID=50379.
10. Angle J (2006) (received 8/05; electronic publication: 12/05; hardcopy publication 7/06). The Inequality Process as a wealth maximizing algorithm. Physica A 367:388–414.
11. Beach C, Slotsve G (1996) Are we becoming two societies? Income Polarization and the Myth of the Declining Middle Class in Canada. Ottawa, Ontario: C.D. Howe Institute.
12. Beach C, Chaykowski R, Slotsve G (1997). Inequality and polarization of male earnings in the United States, 1968–1990. North Am. J. Econ. Fin. 8(2):135–152.
13. Blackburn M, Bloom D (1985) What is happening to the middle class?. Am. Demographics 7(1):18–25.
14. Blackburn M, Bloom D (1987) Earnings and income inequality in the United States. Population and Development Review 13:575–609.
15. Bradbury K (1986) The Shrinking Middle Class. New England Economic Review, September/October, pp. 41–54.

16. Burkhauser R, Crews Cutts A, Daly M, Jenkins S (1999) Testing the significance of income distribution changes over the 1980's business cycle: a cross-national comparison. J. Appl. Econometrics 14(3):253–272.
17. Chakraborti A, Chakrabarti BK (2000) Statistical mechanics of money: how saving propensity affects its distribution. Eur. Phys. J. B 17: 167–170.
18. Chatterjee A, Chakrabarti BK, Manna SS (2003) Money in gas-like markets: Gibbs and Pareto Laws. Phys. Scripta T 106: 36–39.
19. Chatterjee A, Chakrabarti BK, Manna SS (2004) Pareto Law in a kinetic model of market with random saving propensity Physica A 335: 155–163.
20. Chatterjee A, Chakrabarti BK, Stinchcombe RB (2005) Master equation for a kinetic model of trading market and its analytic solution Phys. Rev. E 72: 026126.
21. Chatterjee A, Chakrabarti BK, Stinchcombe RB (2005) Analyzing money distributions in 'ideal gas' models of markets. Practical Fruits of Econophysics, Ed. Takayasu H, pp 333–338, Springer-Verlag, Tokyo.
22. Coder J, Rainwater L, Smeeding T (1989) Inequality among children and elderly in ten modern nations: the United States in an international context. Am. Econ. Review 79(2): 320–324.
23. Current Population Surveys, March 1962–2004. [machine readable data files]/ conducted by the Bureau of the Census for the Bureau of Labor Statistics. Washington, DC: U.S. Bureau of the Census [producer and distributor], 1962–2004. Santa Monica, CA: Unicon Research Corporation [producer and distributor of CPS Utilities], 2004.
24. Duclos J, Esteban J, Ray D (2004) Polarization: concepts, measurement, estimation. Econometrica 72(6):1737–1772.
25. Duncan G, Smeeding T, Rodgers W (1993) Why is the middle class shrinking?. In Papadimitriou D. (ed.), Poverty and Prosperity in the U.S. in the Late Twentieth Century. New York,Macmillan.
26. Esteban J, Ray D (1994) On the measurement of polarization. Econometrica 62(4): 819–851.
27. Esteban J, Ray D (1999) Conflict and distribution. J. Econ. Theory 87: 379–415.
28. Horrigan M, Haugen S (1988) The declining middle class thesis: a sensitivity analysis. Monthly Labor Review 111 (May, 1988): 3–13.
29. Jenkins S (1995) Did the middle class shrink during the 1980's: UK evidence from kernel density estimates. Econ. Letters 49(October, #4): 407–413.
30. Kuttner R (1983) The Declining Middle. Atlantic Monthly July:60–72.
31. Lawrence R (1984) Sectoral Shifts in the Size of the Middle Class. Brookings Review 3: 3–11.
32. Levy F, Michel R (1991) The Economic Future of American Families: Income and Wealth Trends. Washington, DC: Urban Institute Press.
33. Levy F, Murnane R (1992) U.S. Earnings Levels and Earnings Inequality: A Review of Recent Trends and Proposed Explanations. Journal of Economic Literature 30 (3): 1333–1381.
34. Morris M, Bernhardt A, Handcock M (1994) Economic inequality: new methods for new trends. Am. Soc. Review 59: 205–219.
35. Roemer M (2000) Assessing the Quality of the March Current Population Survey and the Survey of Income and Program Participation Income Estimates, 1990–1996. Washington, DC: U.S. Census Bureau. [http://www.census.gov/hhes/www/income/assess1.pdf], accessed 11/9/04.

36. Patriarca M, Chakraborti A, Kaski K (2004) Statistical model with a standard Γ distribution. Phys. Rev. E 70: 016104.
37. Patriarca M, Chakraborti A, Kaski K, Germano G (2005) Kinetic theory models for the distirbution of wealth. arXiv:physics/0504153.
38. Thurow L (1984) The disappearance of the middle class. New York Times Vol.133 (February 5):F3.
39. Unicon Research, inc. 2004. Manual for March Current Population Surveys. Santa Monica, CA: Unicon.
40. U.S. Bureau of Economic Analysis. 2005. National Income and Product Account Table 2.4.4, Price indexes for personal conumption expenditure by type of product [index numbers, 2000 = 100]. http://www.bea.gov/bea/dn/nipaweb/TableView.asp#Mid [Last revised on August 4, 2005].
41. Weinberg D, Nelson C, Roemer M, Welniak E (1999) Fifty years of U.S. income data from the Current Population Survey. Am. Econ. Review 89 (#2) (Papers and Proceedings of the 111th Annual Meeting of the Am. Econ. Assoc.), 18–22.
42. Wolfson M (1994) When inequalities diverge. Am. Econ. Review 84(#2) : 353–358.

Is Inequality Inevitable in Society?
Income Distribution as a Consequence
of Resource Flow in Hierarchical Organizations

Sitabhra Sinha and Nisheeth Srivastava

The Institute of Mathematical Sciences, C.I.T. Campus, Taramani,
Chennai – 600 113, India
sitabhra@imsc.res.in

society (*noun*), from Latin *societas*, equiv. to *soci(us)* partner, comrade + -*etas*, var. of -*itas*- -ity
1. an organized group of persons associated together for religious, benevolent, cultural, scientific, political, patriotic, or other purposes.
2. a body of individuals living as members of a community.
4. a highly structured system of human organization for large-scale community living that normally furnishes protection, continuity, security, and a national identity for its members.

hierarchy (*noun*), from M.L. *hierarchia*, "ranked division of angels"
1. any system of persons or things ranked one above another.

Dictionary.com Unabridged (v 1.1)[1]

Almost all societies, once they attain a certain level of complexity, exhibit inequality in the income of its members. Hierarchical stratification of social classes may be a major contributor to such unequal distribution of income, with intra-class variation often being negligible compared to inter-class differences. In this paper, examples from different historical periods, such as 10th century Byzantium and the Mughal empire of India in the 15th century, and different kinds of organizations, such as a criminal gang in the USA and Manufacturing & IT Services companies in India, are shown to suggest a causal relation between the hierarchical structure of social organization and the observed income inequality in societies. Proceeding from the assumption that income inequality may be a consequence of resource flow in a hierarchically structured social network, we present a model to show that empirically observed long-tailed income distribution can be explained through a process of division of assets at various levels in a hierarchical organization.

---

[1] http://dictionary.reference.com/

1 Introduction

Human society, once it reaches a sufficient degree of complexity, is almost always marked by large inequalities in wealth and income among its members. This is as true of proto-states in the Bronze Age (as indicated by burial remains in the graves of the elite as opposed to common people) as it is in today's highly industrialized world. There have been several attempts at investigating the nature and causes of this pervasive inequality. Vilfredo Pareto (1848–1923) was possibly the first to attempt a quantitative evaluation of social inequality when he collected data on income distribution in several countries including England, Peru, several German states and a number of Italian cities [1]. Pareto claimed that in all cases the data fit a power law relation, with the number of households having income greater than x, $N(x) \sim x^{-\alpha}$. He further made the observation that values of α (now referred to as Pareto exponent) for all the countries observed were around 1.5, possibly the first report of *universality* in a power law relation from empirical data. Based on these observations, Pareto proposed an universal law for income distribution [2]. Later analysis with more accurate and extensive data showed that the power law fits only the long (upper) tail of the income distribution, with the bulk following either a log-normal or Gibbs distribution, and also that the Pareto exponent for different countries have a much wider range of variation than claimed by Pareto [3].

A striking feature of the income distribution predicted by Pareto's law is its extremely skewed nature, with the frequency distribution declining monotonically, beginning at a minimum income [1]. This was referred to by Pareto as the "social pyramid", and brings us to the question of the connection between the observed hierarchical structure of most societies and their income distribution. According to a standard introductory book in prehistoric archaeology, hierarchically organized institutions emerged together with larger, denser populations where decisions could no longer be effectively made by consensus [4]. In turn, hierarchical organizations served to consolidate as well as enhance existing inequalities, so that those having authority remained in power. Thus "civilized" societies are marked by increasing inequality, in wealth as well as in power. Therefore, it seems that the observed long-tailed income distributions can be partially explained to be a result of resource allocation among members belonging to a hierarchical structure, with the minority at the higher levels having a significantly higher income than the majority below. In comparison to inter-level inequality, differences in the income of individuals belonging to the same level, which is due to factors other than the stratified social structure, may be insignificant.

In this paper we present the results of our study involving empirical data analysis and theoretical modeling to show that hierarchical structures can indeed explain the power law form of income distribution in society. In the next section, we analyse the organizational structure and income data from different historical periods, such as that of the Byzantine army in 10th century and the Mughal nobility in 15th century, and from different present-day organizations, such as an urban street gang of Chicago and Indian companies belonging to the Manufacturing and Information Technology sectors. This is followed by a section where we theoretically derive the power law distribution of income from a simple model of resource flow along a hierarchical structure (the tribute model).

2 Empirical Data

2.1 Military Wages in the Byzantine Empire, 10th Century

Between the 10th and 11th centuries, Byzantium was the richest state in the western world, with income levels coming close to the maximum that pre-industrial societies have achieved [5]. We focus on the wages of the Byzantine army, as military expenditure formed the bulk of the total annual budget of the state. This is also seen in its predecessor, the Roman empire, which spent about 80% of its annual budget on the army [6]. The total salaries paid each year to the *strategoi* (generals) of the Byzantine army by the emperor amounted to a total of 26,640 gold coins, which accounted for a significant portion of the state's regular expenditure [7]. Therefore, analyzing the military wages gives a good indication of the overall income distribution in Byzantine society.

Like most military organizations, the Byzantine army had a pyramidal structure, with each theme (a military unit) under the command of a strategos, who was assisted by a hierarchy of lower officers. Even a cursory look at their wages indicates that there was a high degree of inequality. Soldiers were paid between 12–18 nomismata, lower officers between 72–124 nomismata (1–2 pounds of gold), while the strategoi received 1440–2880 nomismata (20–40 pounds of gold), making the top-to-bottom income ratio > 200 [5]. Although the salaries of the strategoi may look extraordinarily high, it included funds to maintain the bodyguard, personal administrative staff and the official residence. It is therefore better to see the *roga* (salary) as an annual budget for running the administration, which included the salary of the official himself [7].

Figure 1 (left) shows the cumulative frequency distribution of wages for soldiers and officers of the *tagmata*, or central fleet, in 911 [8]. The distribution fits a power-law form with a Pareto exponent of $\alpha \sim 2.17$. The deviation of the income for the strategos, which appears too high compared to the general trend of wages for the other officers, is because it presumably includes the money earmarked for administrative expenses as mentioned above. At the lower end, the soldiers were partly paid in kind (grain, clothes, etc) and this could account for the slight deviation.

It is interesting to see that Byzantine army showed both hierarchical structure and the long-tailed income distribution that we associated with Pareto

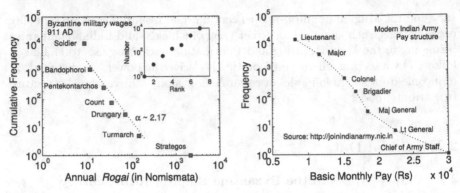

Fig. 1. (*left*) Cumulative frequency distribution of the annual military wages in the Byzantine army in 911 AD. The *dotted line* indicates a power-law fit to the data, with Pareto exponent $\alpha \sim 2.17$. The monetary unit is 1 nomisma, a coin of pure gold weighing 4.5 grams. The *inset* shows the approximate number of people at each rank in a theme (a Byzantine military unit), starting from the strategos. (*Right*) The frequency distribution of basic monthly pay (in Rupees) in the Indian army at present.

law. However, it is not necessary for the two to always coexist. Although modern armies retain the hierarchical organization of the Byzantine army, the salary structure is much more egalitarian. Figure 1 (right) shows the income distribution in the Indian army at present, which suggests an almost exponential decay, especially at the higher ranks. The data for the basic pay has been obtained from the Indian army web-site [9]. Similar kind of salary structure is seen in almost all governmental agencies, which in general show far less inequality than, e.g., companies belonging to the private sector. Presumably this is because such "non-competitive" organizations have other non-monetary benefits, such as, higher job-security.

2.2 Salaries of Mughal Imperial Officers in India, 15th Century

The Mughal empire of India was one of the largest centralized states of the pre-modern world [10]. Between 1556 and 1720 the empire was a large complex organization dominating most of the Indian subcontinent, with the emperor's orders being carried out in every province by a well-trained cadre of military-civilian bureaucrats. The centralized hierarchical administration was built largely during the reign of Emperor Akbar (1556–1605). The reigns of Akbar's successors, his son Jahangir (1605–1627) and grandson, Shah Jahan (1627–1658), saw steady growth in Mughal imperial power and consolidation of the centralized bureaucratic system. The higher administrative positions were filled by a select group of warrior-aristocrats, comprising royal princes and several hundred emirs (nobles). Each of them headed households and troop contingents ranging in size from several hundred to several thousand.

To meet the large expenses involved, the Mughal nobles were paid salaries that were probably the highest in the world at that time [11]. In 1647, during the reign of Emperor Shah Jahan, the 445 highest ranked administrators controlled over 60% of the total imperial revenue income. At the very top, 68 princes and nobles had a combined income of 36.6% of the total revenue, and their *jagirs* covered almost a third of the entire empire [12]. Thus, the income data for the Mughal administrators seems particularly suited for investigating the relation between hierarchical structure and long-tailed income distribution.

With the accession of Emperor Akbar in 1556, the Mughal empire went into a new phase of expansion involving long and costly campaigns. This necessitated streamlining of the army organization and finances, which led Akbar to start in 1577 the *Mansabdari* system (derived from the term *mansab* meaning rank) of hierarchical military nobility. All imperial officers in the civilian and military administration were given ranks, there being 33 grades of mansabdars ranging from commanders of 10 to commanders of 10,000. Each mansabdar was expected to maintain a prescribed number of soldiers, horses, elephants, equipment, etc. according to his *suwar* rank, while the *zat* rank determined his salary. The latter also denoted the standing of the officer in relation to the emperor, the higher the rank the more important the officer. During Akbar's reign, the highest rank an ordinary noble could hold was that of a commander of 5000, the grades between commanders of 7000 and 10,000 being reserved for the royal princes. During the reign of his successors, the grades were increased upto 20,000 or even more [13]. While some officers received their salary directly from the imperial treasury, most were paid through assignment of *jagirs*, the order to collect revenue from certain areas in lieu of salary. Unlike the feudal system in Europe, a mansabdar's property (including jagirs) was not hereditary. Their appointment, promotion or dismissal rested entirely with the emperor.

Detailed information about the salaries of the mansabdars from the middle of Akbar's reign is recorded in the Ain-i Akbari, a detailed statistical account of the empire compiled in 1596–7 by Abu 'l-Fazl (1551–1602), a high-ranking court official [14]. All 249 officers of and above the (zat) rank 500 are mentioned by name in Ain-i Akbari and this includes the three royal princes: Salim (later Emperor Jahangir), Murad and Danyal. In addition, those names of commanders between 500 and 200 who were alive at the time the list was made are also mentioned, while for ranks below 200, only the number of officers who were alive are given. In all, we get information about 1388 mansabdars. This group accounted for the major share of the imperial revenue income: towards the end of Akbar's reign, the mansabdars and their followers consumed 82% of the total annual budget of Rupees 99 million [10].

Figure 2 (left) shows the distribution of rank as well as the monthly salary of the mansabdars mentioned in Ain-i Akbari. The bulk of the officers follow a power law distribution with the Pareto exponent $\alpha \sim 0.69$, with only the royal princes deviating from the general trend. As seen in the inset, the rank

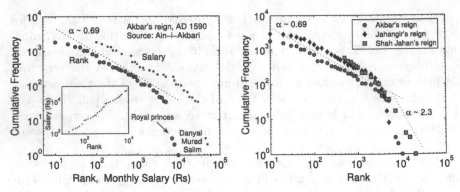

Fig. 2. (*left*) Cumulative frequency distribution of the Mansabdari rank (*filled circles*) and monthly salary in Rupees (*squares*) in the Mughal army around 1590 during the reign of Emperor Akbar. The *line* shows a power-law fit to the data giving a Pareto exponent of $\alpha \sim 0.69$. The *inset* shows the almost linear relation between rank and salary. (*Right*) A comparison of the cumulative frequency distribution of Mansabdari ranks during the successive reigns of Emperors Akbar, Jahangir and Shah Jahan. The data sources are Abu 'l-Fazl's Ain-i Akbari (Akbar), the Dutch traveler De Laët (Jahangir) and Lahori's Padishahnama (Shah Jahan). The latter does not include any information about officers below the rank of 500.

and the salary have an almost linear relationship, which enables us to form an idea of the income distribution in later reigns when sometimes only the rank distribution is available. Figure 2 (right) compares the distribution of mansabdar ranks during the successive reigns of Akbar, Jahangir and Shah Jahan, the bulk of the officers in all cases being described by a Pareto exponent $\sim 0.69^2$. However, as in Akbar's reign, the distribution shows a deviation for the highest ranking officers in the later reigns also, with fewer occupants of these ranks than would be expected from the general trend. This is explained to some extent when we realize that these ranks in almost all cases were reserved for the royal princes. Trying to fit a different power law tail to this part of the data gives us a Pareto exponent of $\alpha \sim 2.3$.

2.3 Income of Members of a Chicago Criminal Gang, 1990s

From a consideration of the hierarchy and income distribution of societies from the long-distant past, we now travel to the present, where a unique study enables us to look at the wages of the members of a criminal organization [15]. The urban street gang which was the subject of this study, a branch of the

---

^{2} It is expected that the nature of the income distribution would also have remain unchanged over this period. Differences would be almost entirely reflected as displacement along the salary axis: because of inflation, the salary of a mansabdar during Shah Jahan's reign was approximately 8 times that of a mansabdar of the same rank during Akbar's reign.

Black Gangster Disciple Nation, operated from Chicago's south side and its principal source of revenue (accounting for more than 75% of the total earnings) was through selling crack cocaine in the street. The street-level salesmen, referred to as foot-soldiers, formed the base of the hierarchical organization of the gang (shown in Fig. 3, left). Other sources of revenue included gang dues from rank-and-file members (full members did not pay dues) and extortion from businesses operating in the gang's territory.

The top level of the organization was made up by a elite group of individuals who were responsible for long-term strategy, maintaining relations with suppliers and affiliates in other regions, and overall gang management [15]. The next level comprised local leaders responsible for specific territories, each of whom were assisted by three "officers", who formed the next lower level. Below them, the foot-soldiers were organized in teams of six, with a team leader, carrier, lookout, etc. Foot-soldiers were paid a flat wage rather than a wage linked to the sales, with the team leader getting the highest and the lookout the lowest wage in the team. However, the actual wage data was not resolved to this level of detail, with only the total wages paid to the foot soldiers under a local leader having been recorded. Figure 3 (right) shows the distribution of monthly salary at the different levels of the gang. It is worth noting that the distribution is extremely skewed, with the local leaders earning 10–20 times more than the average foot soldier. The insets show the exponentially decreasing number of people at each level, and the exponentially increasing salary, up the hierarchy. The same trend has been observed in a study of the wage structure in a medium-sized US firm in a service industry over the years 1969–1988 [16]. We will see in the next section that such oppositely behaving

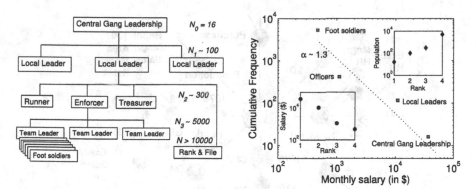

Fig. 3. (*left*) The organizational hierarchy of a Black Disciple Gangster Nation (BDGN) gang in Chicago around the 1990s. The number of gang members at each level is shown on the right. (*Right*) Cumulative frequency distribution of the monthly income for members of the BDGN gang. The *dotted line* is a power law fit with Pareto exponent ~ 1.3. The *upper inset* shows the exponentially increasing number of members at progressively lower levels in the hierarchy, while the *lower inset* shows exponentially decreasing income down the ranks.

exponential functions for the rank-population and rank-salary relations are related to the power law form of the income distribution.

2.4 Management Salary Structure in Indian Companies, 2006

Our final set of empirical data concerns the income of management employees in several Indian companies belonging to the Manufacturing and Information Technology (IT) sectors. Corporate salaries are often quite transparently a manifestation of the hierarchical organization structure of a company. We analysed the salary data from 21 Manufacturing Companies (e.g., Bajaj Auto, Cadbury, Himalaya Drug, etc) and 16 IT Companies (e.g., HCL, Oracle, Satyam, TCS, etc), the bulk of the data being obtained from an internet-based labor market survey site [17], and verified in selected cases with information from the company's annual report.

Figure 4 shows the organization structure of the management in a typical Indian company that was included in our study. The hierarchy starts at the level of executives, and goes up through managers and divisional heads all the way to the Chief Operating Officer or Vice President. The number of levels need not be same in all divisions, e.g., sales and marketing typically has more layers than other divisions, which is reflected in a relatively lower salary ratio between successive levels for people in sales. The salaries at different levels in selected key divisions in companies belonging to the Manufacturing and IT sectors are shown in Fig. 5. The data show exponentially increasing salary along the levels to the top of the hierarchy, a feature already observed in the previous sub-section. Our evidence seems to support that the salary structure

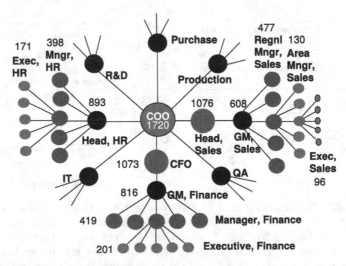

Fig. 4. Organization of the management in a typical Indian company belonging to the manufacturing sector. The numbers correspond to representative annual salaries (in thousands of Rupees) for the respective positions.

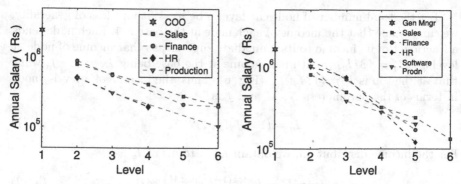

Fig. 5. The management salary structure of (*left*) an Indian company belonging to the manufacturing sector, different from the one whose organization structure is shown in Fig. 4, and (*right*) that of an Indian company belonging to the Information Technology sector, for the same few divisions in each. Note that salaries are shown on a logarithmic scale, so that exponential increase in salary up the hierarchy will be manifested as a linear trend in the figure.

in IT companies is *relatively* more egalitarian than Manufacturing companies, but, on the whole, all companies exhibit similarly skewed salary distribution along their hierarchical structure.

3 Income as Flow along Hierarchical Structure: The Tribute Model

Having seen evidence for a causal relation between long-tailed income distribution and the occurrence of hierarchical structures, we will now consider income distribution as a problem in physics. In particular, we consider income as resource flow along a hierarchically branched structure. We assume the existence of this hierarchy, and observe how power law distribution of resources at the various levels can arise as a result of flow along the structure.

We consider a strict hierarchy of N levels: at each layer, a node is connected to M subordinates belonging to the level below it. The salary of each node is proportional to the total information arriving at it for processing, i.e., the total number of nodes at the level below that it is connected to. Moreover, the income for a node is the difference between the total inflow from the nodes below it and the outflow to the node above it. We shall call this *the tribute model*, as the net flow up the hierarchy can be seen as tribute paid by agents belonging in the lower levels of the hierarchy to those above. An obvious realization of this model is in a criminal organization, where the people at the base put a fraction of their earnings for the disposal of their immediate boss, who in turn sends a fraction to his boss, and so on. We now show that, under certain circumstances, the resulting income distribution will have a power-law form.

Let the total number of nodes at layer q be n_q. Without loss of generality, we can assume that the income of each node at the base is 1. Each node sends a fraction f of its income to its immediate superior, so that income of node at level q is $I_q = fMI_{q+1}$, with net income at the base being $I_N = 1 - f$, while that at the top is $I_1 = (fM)^{N-1}$ (Fig. 6). Thus, the income of a level-q node in terms of the parameters N, M and f is

$$I_q = (1 - f)(fM)^{N-q}. \tag{1}$$

For the income distribution, we obtain n_q in terms of I_q as

$$n_q = n(I) \sim I^{-\log(1-f)/[\log M + \log f]}, \tag{2}$$

which has a power law form. For example, if $f = 1/2$ and $M = 3$, the distribution will be a power law having exponent $\simeq 1.7$. While it may appear that the parameters can be freely chosen to obtain any exponent whatsoever, there are certain restrictions in reality, such as, for a node at the upper layer to benefit from the arrangement, it is necessary that $fM > 1$.

Note that, the salary ratio between two consecutive levels is given by $SR = I_q/I_{q+1} = fM$, so that the top-to-bottom income ratio is

$$T_R = \frac{(fM)^N}{1 - f} = \frac{(SR)^N}{1 - (SR/M)}. \tag{3}$$

For example, if $N = 5$, $M = 5$ and $f = 1/2$, we obtain $T_R = 200$. If, as is empirically observed, T_R remains fairly constant for hierarchical organizations

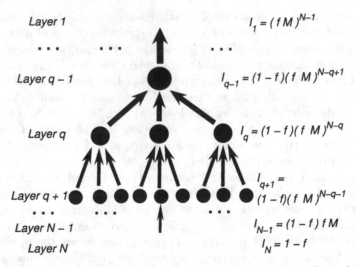

Fig. 6. A schematic diagram of the tribute model representing flow of resources along a hierarchical structure, with the income of nodes at each level q shown in the *right*.

with different number of levels, then it follows that as the number of levels, N, increases, the salary ratio between successive levels ($= fM$) decreases. This has indeed been observed by us in the data for management pay structure in Indian companies.

So far we had been concerned with inter-level differences in income. Within a level also, incomes for individuals will differ. However, as suggested by Gibrat [18], this is likely to be decided by large number of independent stochastic factors, such that the intra-level income distribution will most likely be log-normal, the outcome of a multiplicative stochastic process. When seen in conjunction with the multiple hierarchical levels which ensures that the gap between mean income at each level have exponentially increasing separation along with exponentially decreasing population at each level up the hierarchy, this will imply that within a given level the income distribution is log-normal, but a Pareto like power law behavior will describe the overall inequality, as inter-level differences will tend to dominate intra-level deviations within an organization.

4 Conclusions

In this paper, we have examined the co-occurrence of hierarchical social structures and long-tailed income distributions in many empirical examples, which suggest a causal relation between the two. By considering income distribution as a resource flow problem along a hierarchical structure, we see that it follows a power law form with Pareto exponents similar to those seen in reality. Although factors other than social stratification do play a role in deciding income of individuals, it may be that the broader features of long-tailed income distributions, including the Pareto law behavior at the upper tail, is explained in terms of organizational hierarchy. A future challenge lies in explaining why such hierarchical structures emerge spontaneously in society [19].

Acknowledgement. We thank R.K. Pan for helpful suggestions. SS would like to thank M. Sanyal of the Jadunath Sarkar Bhavan Library, Centre for Studies in Social Sciences, Kolkata for providing invaluable assistance in obtaining the income data for Mughal mansabdars.

References

1. Persky J (1992) Pareto's Law, J. Economic Perspectives 6: 181–192
2. Pareto V (1897) Cours d'Economique Politique, vol 2. Macmillan, London
3. Chatterjee A, Sinha S, Chakrabarti B K (2007) Economic inequality: Is it natural? E-print physics/0703201
4. Price T D, Feinman G M (1997) Images of the Past, 2nd edition. Mayfield Publishing, Mountain View, Calif.

5. Milanovic B (2006) An estimate of average income and inequality in Byzantium around year 1000, Review of Income & Wealth 52: 449–470
6. Duncan-Jones R (1994) Money and Government in the Roman Empire. Cambridge Univ Press, Cambridge
7. Oikonomides N (2002) The role of the Byzantine state in the economy. In: Laiou A E (ed) The Economic History of Byzantium. Harvard Univ Press, Cambridge, Mass.
8. Morrisson C, Cheynet J-C (2002) Prices and wages in the Byzantine world. In: Laiou A E (ed) The Economic History of Byzantium. Harvard Univ Press, Cambridge, Mass.
9. http://joinindianarmy.nic.in/nda13.htm
10. Richards J F (1993) The Mughal Empire. Cambridge Univ Press, Cambridge
11. Chandra S (1982) Standard of living: Mughal India. In: Raychaudhuri T, Habib I (eds) The Cambridge Economic History of India, Vol 1. Cambridge Univ Press, Cambridge
12. Habib I (1974) The social distribution of landed property in pre-British India: A historical survey. In: Sharma R S, Jha V (eds) Indian Society: Historical Probings. Peoples Publishing House, New Delhi
13. Mansabdar, entry in Banglapedia: National Encyclopedia of Bangladesh (http://banglapedia.net/)
14. Blochmann H (1927) Translation of Ain-i Akbari by Abu 'l-fazl Allami, Vol 1, 2nd edition (ed. D C Phillott). Asiatic Society, Calcutta
15. Venkatesh S A, Levitt S D (2000) The financial activities of an urban street gang, Qtrly. J. Econ. 115:755–789
16. Baker G, Gibbs M, Holmstrom B (1994) The internal economics of the firm: Evidence from personnel data, Qtrly. J. Econ. 109: 881–919
17. http://www.paycheck.in/
18. Gibrat R (1931) Les Inégalites Économiques. Librairie du Recueil Sirey, Paris
19. Watts D J, Dodds P S, Newman M E J (2002) Identity and search in social networks, Science 296: 1302–1305

Knowledge Sharing and R&D Investment

Abhirup Sarkar

Economic Research Unit, Indian Statistical Institute, Kolkata
abhirup@isical.ac.in

Summary. We consider an R&D race between two symmetric firms. The game consists of two stages. In the first stage, firms non-cooperatively decide upon their levels of investment in R&D which, in turn, determine the Poisson probabilities of their subsequent sucesses. In the second stage, they engage in a Nash bargaining game to share their knowledge. We show that the firms over-invest and earn lower profits if knowledge sharing is possible compared to the situation where it is not. Hence, before the first stage, if the firms are given the option of precommitting to no knowledge sharing, they will do so and be better off. The society, of course, would be better off with full knowledge sharing.

1 Introduction

In recent years, a lot of interest has been created on the process of knowledge accumulation by firms through R&D investments. A part of this knowledge accumulation process consists of knowledge sharing by firms. Often production and sales of a product is preceded by a long process of research. This is particularly true in case of new product development. Typically, the R&D stage can be broken down into several sub-stages and in some of these intermediate stages firms may want to share their knowledge with one another. More specifically, the firm with the superior knowledge may sell its information to firms with inferior knowledge for a licensing fee. Due to the public good nature of knowledge, such sharing would always be welfare improving for the society. But the question is do the firms have an incentive to share their knowledge? Most of the existing literature concentrates on bargaining mechanisms through which knowledge sharing might take place between firms. Thus Green and Scotchmar (1995), Aghion and Tirole (1994), dealing with multistage patent races and with R&D knowledge selling arrangements respectively, analyse the problem of dividing the expected surplus from ultimate R&D success among generators and utilizers of disclosed interim research knowledge. In these papers, it is assumed that the generator of knowledge through basic research is unable to develop it further into a marketable invention and the agent who is able to market the knowledge in incapable of doing basic research. In a more recent paper, d'Aspremont, Bhattacharya and Gerard-Varet (2000) consider a framework of knowledge-sharing where both the parties can subsequently develop the interim knowledge. In Erkal and Minehart (2006), dynamic aspects of knowledge sharing in R&D rivalry are considered. In particular, it focuses on a situation where research projects consist of a number of stages and explores how the innovators incentives to share intermediate research outcomes with progress and with their relative positions in the R&D race. In contrast, the present paper looks at the desirability of knowledge sharing from the point of view of the firms.It adds a prior stage to the R&D race where firms decide non-cooperatively the level of investment to be made on research which subsequently determines the probabilities of their ultimate success. We show that in such a scenario firms tend to overinvest in R&D and their profits are lower than in the situation where there is no possibility of knowledge sharing. Thus if the firms could precommit to no knowledge sharing, they would do so. The rate of R&D would, however, be much higher with knowledge sharing because it will increase the probability of success in R&D.

2 The Basic Model

We consider two symmetric firms engaged in an R&D race. There are two stages of this game. In the first stage, each firm chooses its level of investment in R&D which determines its Poisson intensity of invention. Thus if the ith firm invests $C(\lambda_i)$ on R&D in the first stage, it has a Poisson intensity of invention λ_i. We assume that the cost function $C(\lambda_i)$ is the same for both the firms and $C' > 0$, $C'' > 0$. In the second stage, the firms are engaged in a (symmetric) Nash bargaining game of knowledge sharing. As an outcome of the game, the firm with superior knowledge transfers its information to the firm with inferior knowledge against a fixed licensing fee.

Let us first consider a benchmark case where no disclosure of knowledge is possible at the second stage of the game. Suppose both firms share a common discount rate δ in continuous time. We further assume that for each firm the success in innovation is a random event governed by a Poisson process. Let T_i be the time when success in innovation occurs for the ith firm. Then the distribution of T_i is given by

$$F(T_i) = 1 - e^{-\lambda_i T_i} \tag{1}$$

where $F(T_i)$ measures probability that success occurs on or before date T_i. Hence the probability density of T_i is given by

$$f(T_i) = F'(T_i) = \lambda_i e^{-\lambda_i T_i} dT_i \tag{2}$$

Thus the probability that innovation will occur sometime within the short interval between T_i and $T + dT_i$ is approximately $\lambda_i e^{-\lambda_i T} dT_i$. In particular, the probability that it will occur within dT_i from *now*, i.e. when $T_i = 0$, is approximately $\lambda_i dT_i$. So we can interpret λ_i as the probability *per unit of time* that innovation will occur. We may call this the flow probability of the event. We assume that innovation by the two firms are governed by two similar but independent Poisson proceses. Hence the flow probability that at least one firm will be successful in innovation is just the sum $\lambda_1 + \lambda_2$ and the conditional probability that firm 1, with Poisson intensity of invention λ_1, is the first innovator is given by $\frac{\lambda_1}{\lambda_1 + \lambda_2}$. Finally, we assume that the first innovator gets all the benefits of innovation, which is normalized to unity, and the other firm gets nothing. Then the discounted expected payoff of firm 1 in the no disclosure scenario is given by

$$\pi_1^N(\lambda_1, \lambda_2) = \int_0^\infty e^{-\delta t} \frac{\lambda_1}{\lambda_1 + \lambda_2} f(t) dt = \frac{\lambda_1}{\lambda_1 + \lambda_2 + \delta} \tag{3}$$

where $f(t)$ represents the probability that at least one firm innovates within the short time interval $t + dt$. In particular, remaining consistent with equation (2) above $f(t)$ is given by

$$f(t) = (\lambda_1 + \lambda_2) e^{-(\lambda_1 + \lambda_2)t} \tag{4}$$

Similarly the discounted expected payoff of firm 2 in the no disclosure scenario is given by

$$\pi_2^N(\lambda_1, \lambda_2) = \frac{\lambda_2}{\lambda_1 + \lambda_2 + \delta} \tag{5}$$

Let us now consider the case where in the second stage knowledge is transfered from the informed to the uninformed firm. Suppose firm 1 has the superior knowledge, i.e. $\lambda_1 \geq \lambda_2$. then the payoff function of firm 1 is given by

$$\pi_1(\lambda_1, \lambda_2) = \frac{\lambda_1}{2\lambda_1 + \delta} + L(\lambda_1, \lambda_2) \tag{6}$$

and that of firm 2 is given by

$$\pi_2(\lambda_1, \lambda_2) = \frac{\lambda_1}{2\lambda_1 + \delta} - L(\lambda_1, \lambda_2) \tag{7}$$

In the above two equations $L(\lambda_1, \lambda_2)$ represents the licensing fee paid by firm 2 to firm 1 when knowledge is transfered. We assume that the level of knowledge is verifiable by both parties. Also we assume for the moment that there is full disclosure of knowledge by firm 1 to firm 2. However, we shall show below that firm 1 will disclose its entire knowledge to firm 2 even if it has the option of partial disclosure. The licensing fee is determined by a symmetric Nash bargaining solution. In other words, the licensing fee satifies

$$\pi_i = \pi_i^N(\lambda_1, \lambda_2) + \frac{1}{2} \left[\pi_1(\lambda_1, \lambda_2) + \pi_2(\lambda_1, \lambda_2) - \pi_1^N(\lambda_1, \lambda_2) - \pi_2^N(\lambda_1, \lambda_2) \right] \tag{8}$$

This means that the net surplus from knowledge sharing is equally divided between the two firms. Using equations (6) and (7), the full disclosure payoffs may be written as

$$\pi_i(\lambda_1, \lambda_2) = \frac{\lambda_i}{\lambda_1 + \lambda_2 + \delta} + \frac{1}{2}\left[\frac{2\lambda_1}{2\lambda_1 + \delta} - \frac{\lambda_1 + \lambda_2}{\lambda_1 + \lambda_2 + \delta}\right] \tag{9}$$

for $i = 1, 2$. It is to be noted that the net surplus from knowledge sharing is non-negetive due to the public good nature of knowledge. We know consider the first stage of the game where the firms decide upon their R&D expenditure which determines their levels of knowledge. When there is no disclosure in stage 2, the net profit functions may be written as

$$\Pi_i^N(\lambda_1, \lambda_2) = \frac{\lambda_i}{\lambda_1 + \lambda_2 + \delta} - C(\lambda_i) \tag{10}$$

for $i = 1, 2$. Similarly, when there is full disclosure, the profit functions may be written as

$$\Pi_i^F(\lambda_1, \lambda_2) = \frac{\lambda_i}{\lambda_1 + \lambda_2 + \delta} + \frac{1}{2}\left[\frac{2\lambda_1}{2\lambda_1 + \delta} - \frac{\lambda_1 + \lambda_2}{\lambda_1 + \lambda_2 + \delta}\right] - C(\lambda_i) \tag{11}$$

for $i = 1, 2$ and where $\lambda_1 \geq \lambda_2$. We compare the Nash equilibrium of equation (9) with that of (10). The first order conditions of the equilibrium with no disclosure are given by

$$\frac{\lambda_2 + \delta}{(\lambda_1 + \lambda_2 + \delta)^2} = C'(\lambda_1) \tag{12}$$

$$\frac{\lambda_1 + \delta}{(\lambda_1 + \lambda_2 + \delta)^2} = C'(\lambda_2) \tag{13}$$

Since $C''(.) > 0$, in equilibrium the two firms must choose the same level of R&D investment. This equilibrium is denoted by $\lambda_1^N = \lambda_2^N = \lambda^N$. Similarly the first order conditions for the case where there is full disclosure of knowledge are given by

$$\frac{\lambda_2 + \delta}{(\lambda_1 + \lambda_2 + \delta)^2} + \frac{\delta}{2}\left[\frac{2}{(2\lambda_1 + \delta)^2} - \frac{1}{(\lambda_1 + \lambda_2 + \delta)^2}\right] = C'(\lambda_1) \tag{14}$$

$$\frac{\lambda_1 + \delta}{(\lambda_1 + \lambda_2 + \delta)^2} + \frac{\delta}{2}\left[\frac{2}{(2\lambda_1 + \delta)^2} - \frac{1}{(\lambda_1 + \lambda_2 + \delta)^2}\right] = C'(\lambda_2) \tag{15}$$

It can be shown that in this case also $\lambda_1 = \lambda_2$ in Nash equilibrium. To see why, first note that for $\lambda_1 = \lambda_2$ the left hand sides of equations (13) and (14) become identical, so that $\lambda_1 = \lambda_2$ is clearly a solution to (13) and (14). Next, we may show that this is the only solution. If possible, suppose $\lambda_1 > \lambda_2$. On the one hand, in view of the fact that $C'' > 0$, this means that the right hand side of equation (13) is greater than the right hand side of equation (14). On the other hand, it is straight forward to verify that the left hand side of equation (13) is less than the left hand side of equation (14). This is a contradiction which proves that $\lambda_1 > \lambda_2$ is not possible. Thus we conclude that in case of full disclosure, levels of investment and knowledge are the same for the two firms and this common knowledge is denoted by $\lambda_1^F = \lambda_2^F = \lambda^F$.

Equilibrium R&D in the no disclosure and full disclosure cases are given by

$$\frac{\lambda^N + \delta}{(2\lambda^N + \delta)^2} = C'(\lambda^N) \tag{16}$$

$$\frac{\lambda^F + \delta}{(2\lambda^F + \delta)^2} + \frac{\delta}{2}\frac{1}{(2\lambda^F + \delta)^2} = C'(\lambda^F) \tag{17}$$

respectively. It is straight forward to verify from equations (15) and (16) that $\lambda^F > \lambda^N$. This has direct implications for the comparative profit levels of the firms in the two situations.

Note that the level of profit (of each firm) in the two situations is given by $\frac{\lambda^j}{2\lambda^j + \delta} - C(\lambda^j), j = N, F$. Also note that the function $\frac{\lambda}{2\lambda + \delta} - C(\lambda)$ attains a maximum where $\frac{\delta}{(2\lambda + \delta)^2} = C'(\lambda)$. Denoting the value of λ where this is satisfied, by λ^{max} it is easy to see that $\lambda^F > \lambda^N > \lambda^{max}$. It follows immediately that profit of each firm will unambiguously go down if they are allowed to share knowledge in the second stage. The level of R&D, however, is higher under knowledge-sharing than under no disclosure. A couple of comments are now in order. First, in the second stage of the game it is always optimal for the firms to go for knowledge-sharing. The firms will take this into account and as a result invest more in R&D in the first stage leading to lower profits. Secondly, it may be pointed out that in the second stage a superior firm (provided there is one) will always go for full knowledge sharing. Suppose in the second stage firm 1 has superior knowledge so that $\lambda_1 > \lambda_2$. Then if firm 1 reveals a level of knowledge $\tilde{\lambda}$ to firm 2 with $\lambda_1 > \tilde{\lambda} > \lambda_2$, its payoff is given by

$$\pi_1(\lambda_1, \lambda_2, \tilde{\lambda}) = \frac{2\lambda_1 + \tilde{\lambda} + \lambda_2}{2(\lambda_1 + \lambda_2 + \delta)} \tag{18}$$

Since the right hand side of (17) is increasing in $\tilde{\lambda}$, it is always optimal for firm 1 to reveal full knowledge, that is λ_1', to firm 2. The above findings are summarized in the following proposition.

Proposition 1 *In the R&D race described above, if knowledge-sharing is allowed, then each firm will be making lower profits compared to the case where knowledge-sharing is not possible. The aggregate level of R&D, however, will be higher under knowledge-sharing. In equilibrium there will be no knowledge sharing.*

3 Concluding Remarks

The paper shows, in the context of an R&D race, that firms will over invest in research if they have the opportunity of sharing their knowledge at a subsequent stage. This will also reduce their profits. Therefore, if a proper market for knowledge sharing is not in place, firms will have little incentives to develop them on their own. But knowledge sharing can potentially increase the possibility of innovation in a couple of ways. First, by sharing knowledge all firms have the same probability of success which is equal to the probability of success of the informationally most superior firm. This, of course, increases the probability that at least one firm is successful within any given time interval. Secondly, due to over investment in research, the probability of success of the superior firm will be higher. In our model of symmetric firms, since in equilibrium each firm will be investing the same amount, each will have the same level of knowledge at the intermediate stage. Thus in equilibrium there will be no knowledge sharing. But the mere possibility of knowledge sharing will increase investment in R&D which will certainly be beneficial for the society. Thus if there is no well developed institution through which knowledge can be shared, the government will have to take the initiative to develop such institutions or arrangements. More so, because the firms will not have any incentive to develop them.

References

1. Aghion P, Tirole J (1994) On the Management of Innovation, Quarterly Journal of Economics 109: 1185–1209.
2. d'Aspremont C, Bhattacharya S, Gerard-Varet L (2000) Bargaining and Sharing Innovative Knowledge. Review of Economic Studies 67: 255–271.
3. Erkal N, Minehart D (2005) Optimal Sharing Strategies in Dynamic Games of Research and Development, Working Paper, University of Melbourne, Victoria, Australia.
4. Green J, Scotchmer S (1995) Antitrust Policy, the Breadth of Patent Protection and the Incentive to Develop New Products, Rand Journal of Economics 26: 20–33.

Preferences Lower Bound in the Queueing Model

Manipushpak Mitra

Economic Research Unit, Indian Statistical Institute, Kolkata
mmitra@isical.ac.in

Summary. We show the existence of a first best incentive compatible mechanism for the queueing model that satisfies identical preferences lower bound. We call this mechanism the FB' mechanism. We also show that for the queueing model, either with three agents or with four agents, the FB' mechanism is the only first best incentive compatible mechanism that satisfies identical preferences lower bound.

1 Introduction

In a queueing environment, we have a finite set of agents who wish to avail a service provided by one server. Agents can be served sequentially. Agents incur waiting costs and the planner's goal is efficiency of decision, that is, to order the agents in a queue to minimize the aggregate waiting costs. Maniquet [4] argues that queueing models capture many economic environments. Such models have been examined from both incentive and axiomatic viewpoints in a recent series of papers. The problem from an incentive viewpoint occurs if agents have private information. In such a scenario, individual objective differs from the planner's objective. If waiting costs are linear functions of time and agents have quasi-linear preferences, then one can find first best incentive compatible mechanisms. By first best incentive compatible mechanisms we mean mechanisms that satisfy truth-telling in dominant strategies, efficiency of decision and budget balanceness.[1]

One question that can be asked is the following: can we have first best incentive compatible mechanism that results in utility allocation that satisfies some reasonable distributional requirements. An interesting distributional requirement is that each agent should find his bundle (that is, the allocated queue position and transfer under the mechanism) at least as desirable as the bundle that would be assigned to him in the hypothetical situation in which all agents would have preferences identical to his (see Thomson [8]).

---

[1] See Mitra [5].

If a mechanism satisfies this distributional requirement then we say that the mechanism satisfies identical preferences lower bound. This concept was proposed by Moulin [6] and has been the object of several studies in the context of economies with indivisible goods (see Bevia [1], [2]). In modeling queueing situations as a coalitional form TU game, both Maniquet [4] and Chun [3] has used identical preferences lower bound to characterize the Shapley value (see Shapley [7]) of their respective queueing games. We show the existence of a first best incentive compatible mechanism that satisfies identical preferences lower bound. We call this mechanism the FB' mechanism. We also show that for the queueing model, either with three agents or with four agents, the FB' mechanism is the only first best incentive compatible mechanism that also satisfies identical preferences lower bound.

2 The Model

Let $N = \{1, \ldots, n\}$ be the set of agents and assume that $n \geq 3$. Each agent wants to consume a service provided by a server. Agents can be served only sequentially and serving any agent takes one unit of time. Each agent is identified with a waiting cost $\theta_i \in \Re_+$. The waiting cost is private information. A state $\theta = (\theta_i)_{i \in N}$ is a vector of waiting costs of all agents. Given a state θ, the state $(\theta_j)_{j \in N - \{i\}}$ will be denoted θ_{-i}.

A queue is a bijection $\sigma : N \to \{1, \ldots, |N|\}$. For notational convenience, we will denote $\sigma(i)$ as σ_i for all $i \in N$. If $\sigma_i = k$ – that is, agent i occupies position k in the queue – then her waiting cost is $(k-1)\theta_i$.

2.1 Mechanism Design Problem

If the waiting cost of the agents are private information we have a mechanism design problem of the planner. A mechanism $\mathcal{M} = (\sigma, t)$ associates to each state θ, a tuple $(\sigma(\theta), t(\theta)) \in \Sigma(N) \times \Re^N$. Agent i's outcome is denoted as $(\sigma_i(\theta), t_i(\theta))$ where $\sigma_i(\theta)$ is the queue position of agent i and $t_i(\theta)$ the corresponding transfer. Let $u_i(\sigma_i(\theta), t_i(\theta), \theta_i') = -(\sigma_i(\theta) - 1)\theta_i' + t_i(\theta)$ denote i's utility when the state θ is reported (collectively) and her true waiting cost is θ_i'.

Definition 1 A mechanism (σ, t) is *dominant strategy incentive compatible* if for all $i, \theta_i, \theta_i', \theta_{-i}$, $u_i(\sigma_i(\theta_i, \theta_{-i}), t_i(\theta_i, \theta_{-i}), \theta_i) \geq u_i(\sigma_i(\theta_i', \theta_{-i}), t_i(\theta_i', \theta_{-i}), \theta_i)$.

Dominant strategy incentive compatibility implies that truth telling is an optimal strategy for every agent irrespective of the announcements of other agents.

Definition 2 A mechanism (σ, t) is *efficient* if for all states θ, $\sigma(\theta) \in \Sigma^*(N, \theta)$.

It is easy to see that efficiency implies that if $\theta_i > \theta_j$ then $\sigma_i < \sigma_j$. However, if $\theta_i = \theta_j$, then the efficient queue can involve either $\sigma_i > \sigma_j$ or $\sigma_i < \sigma_j$. In this paper, we avoid this multiplicity problem by focusing on a single valued selection from the efficiency correspondence. This is done by selecting an arbitrary order $\succ$ on the set of agents N and using the following tie breaking rule: if $i \succ j$ and $\theta_i = \theta_j$ then $\sigma_i < \sigma_j$. In what follows, we will use $\sigma^*(\theta)$ to denote the efficient queue consistent with the above tie-breaking rule for the state θ.

Definition 3 A mechanism (σ, t) is *budget balanced* if for all states θ, $\sum_{i=1}^n t_i(\theta) = 0$.

A mechanism that satisfies efficiency, dominant strategy incentive compatibility and budget balance is called a *first best incentive compatible mechanism*.

Proposition 2 A mechanism $\mathcal{M} = (\sigma^*, t)$ is first best incentive compatible if and only if for all $\theta \in \Re_+^n$ and all $i \in N$,

$$t_i(\theta) = \sum_{j \in P_i(\sigma^*(\theta))} \left(\frac{\sigma_j^*(\theta) - 1}{n - 2} \right) \theta_j - \sum_{j \in P_i'(\sigma^*(\theta))} \left(\frac{n - \sigma_j^*(\theta)}{n - 2} \right) \theta_j + \gamma_i(\theta_{-i}) \quad (1)$$

where $P_i(\sigma^*(\theta)) = \{j \in N \mid \sigma_j^*(\theta) < \sigma_i^*(\theta)\}$, $P_i'(\sigma^*(\theta)) = \{j \in N \mid \sigma_j^*(\theta) > \sigma_i^*(\theta)\}$, $\gamma_i : \Re_+^{n-1} \to \Re$ and $\sum_{i \in N} \gamma_i(\theta_{-i}) = 0.$[2]

Definition 4 The FB$'$ mechanism, denoted by $\mathcal{M}' = (\sigma^*, t')$, is a first best incentive compatible mechanism with the following transfer:

$$t_i'(\theta) = \sum_{j \in P_i(\sigma^*(\theta))} \left(\frac{\sigma_j^*(\theta) - 1}{n - 2} \right) \theta_j - \sum_{j \in P_i'(\sigma^*(\theta))} \left(\frac{n - \sigma_j^*(\theta)}{n - 2} \right) \theta_j \quad (2)$$

for all $\theta \in \Re_+^n$ and all $i \in N$.

Clearly, the FB$'$ mechanism is a first best incentive compatible mechanism with $\gamma_i(\theta_{-i}) = 0$ for all $i \in N$ and all $\theta_{-i} \in \Re_+^{n-1}$. We denote the utility allocation, associated with the FB$'$ mechanism, in any state θ, by $u'(\theta) = (u_1'(\theta), \ldots, u_n'(\theta))$ where

$$u_i'(\theta) \equiv u_i(\sigma_i^*(\theta), t_i'(\theta), \theta_i) = -(\sigma_i^*(\theta) - 1)\theta_i + t_i'(\theta) \quad (3)$$

2.2 Identical Preferences Lower Bound

Definition 5 A mechanism $\mathcal{M} = (\sigma, t)$ satisfies *identical preferences lower bound* if for all $\theta \in \Re_+^n$, $u_i(\sigma_i(\theta), t_i(\theta), \theta_i) \geq -\frac{(n-1)\theta_i}{2}$ for all $i \in N$.

---

[2] For the proof of this result see Mitra [5].

The property of identical preferences lower bound requires that each agent should be at least as well off as he would be if all other agents had the same preferences.

Proposition 3 The FB′ mechanism satisfies identical preferences lower bound.

Proof: To show that the FB′ mechanism satisfies identical preferences lower bound we first substitute (2) in (3) to obtain

$$u_i'(\theta) = -(\sigma_i^*(\theta) - 1)\theta_i + \sum_{j \in P_i(\sigma^*(\theta))} \beta_j(\theta)\theta_j - \sum_{j \in P_i'(\sigma^*(\theta))} \tilde{\beta}_j(\theta)\theta_j \quad (4)$$

where $\beta_j(\theta) = \left(\frac{\sigma_j^*(\theta)-1}{n-2}\right)$ and $\tilde{\beta}_j(\theta) = \left(\frac{n-\sigma_j^*(\theta)}{n-2}\right)$. We add and subtract $\sum_{j \in P_i(\sigma^*(\theta))} \beta_j(\theta)\theta_i - \sum_{j \in P_i'(\sigma^*(\theta))} \tilde{\beta}_j(\theta)\theta_i$ to the right hand side of (4) and then simplify it to obtain

$$u_i'(\theta) = -\frac{(n-1)\theta_i}{2} + \sum_{j \in P_i(\sigma^*(\theta))} \beta_j(\theta)(\theta_j - \theta_i) + \sum_{j \in P_i'(\sigma^*(\theta))} \tilde{\beta}_j(\theta)(\theta_i - \theta_j) \quad (5)$$

From (5) it follows that $u_i'(\theta) \geq -\frac{(n-1)\theta_i}{2}$ since $\beta_j(\theta) \geq 0, \tilde{\beta}_j(\theta) \geq 0, \theta_j - \theta_i \geq 0$ for all $j \in P_i(\sigma^*(\theta))$ and $\theta_i - \theta_j \geq 0$ for all $j \in P_i'(\sigma^*(\theta))$.

Is FB′ mechanism the only mechanism belonging to the class of first best incentive compatible mechanism that satisfies identical preferences lower bound? The next proposition is a partial answer to this question.

Proposition 4 If $n \in \{3,4\}$, then the FB′ mechanism is the only first best incentive compatible mechanism that satisfies identical preferences lower bound.

Proof: Consider the queueing model with $n = 3$ and a state $\theta = (\theta_1, \theta_2, \theta_3) \in \Re_+^3$ such that $\theta_1 > \theta_2 > \theta_3$. Under any first best incentive compatible mechanism $M = (\sigma^*, t), \sigma_2^*(\theta) = 2$ and agent 2's utility allocation is $U_2(\theta) = -\theta_2 - \gamma_2(\theta_1, \theta_3)$. Applying identical preferences lower bound we get $U_2(\theta) \geq -\theta_2$ which implies that $\gamma_2(\theta_1, \theta_3) \leq 0$. Consider state $\theta' = (\theta_1', \theta_2, \theta_3) \in \Re_+^3$ such that $\theta_2 > \theta_1' > \theta_3$. The state θ' is obtained from the state θ by replacing θ_1 with θ_1'. Under any first best incentive compatible mechanism $M = (\sigma^*, t)$, $\sigma_1^*(\theta') = 2$ and agent 1's utility allocation is $U_1(\theta') = -\theta_1' - \gamma_1(\theta_2, \theta_3)$. Applying identical preferences lower bound we get $U_1(\theta') \geq -\theta_1'$ which implies that $\gamma_1(\theta_2, \theta_3) \leq 0$. Consider state $\theta'' = (\theta_1, \theta_2, \theta_3'') \in \Re_+^3$ such that $\theta_1 > \theta_3'' > \theta_2$. Using similar steps we get $\gamma_3(\theta_1, \theta_2) \leq 0$. Therefore, it is necessary that $\gamma_i(\theta_j, \theta_k) \leq 0$ for all $i \neq j \neq k \neq i$ and all $\theta_j \neq \theta_k$. Given that for a first best incentive compatible mechanism $\sum_{i=1}^3 \gamma_i(\theta_{-i}) = 0$, it follows that **(a)** $\gamma_i(\theta_j, \theta_k) = 0$ for all $i \in \{1, 2, 3\}$, all $\theta_j, \theta_k \in \Re_+$ such that $\theta_j \neq \theta_k$. Finally, if $\bar{\theta} = (\bar{\theta}_1, \bar{\theta}_2, \bar{\theta}_3) \in \Re_+^3$ is such that $\bar{\theta}_1 = \bar{\theta}_2 = \bar{\theta}_3$ then using identical preferences lower bound condition of each agent we get $\gamma_i(\bar{\theta}_j, \bar{\theta}_k) \leq 0$ for all $i \neq j \neq k \neq i$. Given $\sum_{i=1}^{3} \gamma_i(\bar{\theta}_j, \bar{\theta}_k) = 0$, we get **(b)** $\gamma_i(\bar{\theta}_j, \bar{\theta}_k) = 0$ for all $i \in \{1, 2, 3\}$ for all $\bar{\theta}_j = \bar{\theta}_k \in \Re_+$. From **(a)** and **(b)** it follows that $\gamma_i(\theta_{-i}) = 0$ for all $i \in \{1, 2, 3\}$ and for all $\theta_{-i} \in \Re_+^2$.

Consider the queueing model with $n = 4$ and a state $\theta = (\theta_1, \theta_2, \theta_3, \theta_4) \in \Re_+^4$ such that $\theta_1 > \theta_2 > \theta_3 \geq \theta_4$. Under any first best incentive compatible mechanism $M = (\sigma^*, t)$, $\sigma_2^*(\theta) = 2$ and agent 2's utility allocation $U_2(\theta) = -\theta_2 - \frac{1}{2}\theta_3 - \gamma_2(\theta_1, \theta_3, \theta_4)$. Applying identical preferences lower bound we get $U_2(\theta) \geq -\frac{3}{2}\theta_2$ which implies that $\gamma_2(\theta_1, \theta_3, \theta_4) \leq \frac{1}{2}(\theta_2 - \theta_3)$. Notice that the last inequality must be true for any selection of $\theta_2 \in (\theta_3, \theta_1)$ (since $\gamma_2(\theta_1, \theta_3, \theta_4)$ is independent of θ_2). Therefore, $\gamma_2(\theta_1, \theta_3, \theta_4) \leq \frac{1}{2}(\theta_2 - \theta_3)$ must be true in the limit when $\theta_2 = \theta_3$. Hence it is necessary that $\gamma_2(\theta_1, \theta_3, \theta_4) \leq 0$. By applying similar steps one can show that $\gamma_i(\theta_{-i}) \leq 0$ for all $i \in \{1, 2, 3, 4\}$. Given $\sum_{i=1}^{4} \gamma_i(\theta_{-i}) = 0$, we get **(c)** $\gamma_i(\theta_{-i}) = 0$ for all $i \in \{1, 2, 3, 4\}$ and for all $\theta_{-i} \in \Re_+^3$ such that all elements are not equal. If $\bar{\theta} = (\bar{\theta}_1, \bar{\theta}_2, \bar{\theta}_3, \bar{\theta}_4) \in \Re_+^4$ is such that $\bar{\theta}_1 = \bar{\theta}_2 = \bar{\theta}_3 = \bar{\theta}_4$ then by applying identical preferences lower bound for each agent we get $\gamma_i(\bar{\theta}_{-i}) \leq 0$ for all $i \in \{1, 2, 3, 4\}$. Given $\sum_{i=1}^{4} \gamma_i(\bar{\theta}_{-i}) = 0$, we get **(d)** $\gamma_i(\bar{\theta}_j, \bar{\theta}_k, \bar{\theta}_l) = 0$ for all $i \in \{1, 2, 3, 4\}$ and for all $\bar{\theta}_j = \bar{\theta}_k = \bar{\theta}_l \in \Re_+$. From **(c)** and **(d)** we get $\gamma_i(\theta_{-i}) = 0$ for all $i \in \{1, 2, 3, 4\}$ for all $\theta_{-i} \in \Re^3$.

It is not clear whether the uniqueness result of Proposition 4 holds for $n > 4$. An open question in this context would be to axiomatize the allocation under the FB' mechanism using some 'reasonable' fairness axioms.

References

1. Bevia C (1994) Identical preferences lower bound solution and consistency in economies with indivisible goods. Social Choice and Welfare 13: 113–126.
2. Bevia C (1998) Fair allocation in a general model with indivisible objects. Review of Economic Design 3: 195–213.
3. Chun Y (2004) A note on Maniquet's characterization of the Shapley value in queueing problems. Working paper, Rochester University.
4. Maniquet F (2003) A characterization of the Shapley value in queueing problems. Journal of Economic Theory 109: 90–103.
5. Mitra M (2001) Mechanism design in queueing problems. Economic Theory 17: 277–305.
6. Moulin H (1990) Fair division under joint ownership: recent results and open problems. Social Choice and Welfare 7: 149–170.
7. Shapley LS (1953) A value for n-person games. In: Contributions to the Theory of Games II (Eds. Kuhn H, Tucker AW). Princeton University Press, Princeton, 307–317.
8. Thomson W (2003) On monotonicity in economies with indivisible goods. Social Choice and Welfare 21: 195–205.

Kolkata Restaurant Problem
as a Generalised El Farol Bar Problem

Bikas K. Chakrabarti

Theoretical Condensed Matter Physics Division and Centre for Applied
Mathematics and Computational Science, Saha Institute of Nuclear Physics 1/AF
Bidhan Nagar, Kolkata 700064, India
bikask.chakrabarti@saha.ac.in

Summary. Generalisation of the El Farol bar problem to that of many bars here
leads to the Kolkata restaurant problem, where the decision to go to any restaurant
or not is much simpler (depending on the previous experience of course, as in the
El Farol bar problem). This generalised problem can be exactly analysed in some
limiting cases discussed here. The fluctuation in the restaurant service can be shown
to have precisely an inverse cubic behavior, as widely seen in the stock market
fluctuations.

Key words: Stock market fluctuations; El Farol Bar problem; Traffic jam;
Fiber bundle model

1 Introduction

The observed corrlated fluctuations in the stock markets, giving power law
tails for large fluctuations (in contrast to the traditionally assumed exponentially decaying Gaussian fluctuations of random walks), were schematically
incorporated in the El Farol Bar problem of Arthur [1]. Here the decision to
occupy the bar (buy the stock) or to remaing at home (sell the stock) depends
on the previous experience of the "crowd" exceeding the threshold (of pleasure
or demand level) of the bar (stock), and the strategies available. The resulting
Minority Game models [2] still fail to get the ubiquitus inverse cubic law of
stock fluctuations [3]. In the Fiber Bundle models [4] of materials' fracture,
or in the equivalent Traffic Jam models [5], the fibers or the roads fail due to
load, exceeding the (preassigned) random threshold, and the extra load gets
redistributed in the surviving fibers or roads; thereby inducing the corelations
in the fluctuations or "avalanche" sizes. The avalanche distribution has a clear
inverse cubic power law tail in the "equal load sharing" or "democratic" fiber
bundle model [6, 7].

   In the El Farol Bar problem [1], the Santa Fe people decide whether to
go to the bar this evening, based on his/her experince last evening(s). The

bar can roughly accommodate half the poulation of the (100-member strong) Institute and the people coming to the bar still enjoy the music and the food. If the crowd level goes beyond this level, people do not enjoy and each of those who came to the bar thinks that they would have done better if they stayed back at home! Clearly, people do not randomly choose to come to the bar or stay at home (as assumed in a random walk model); they exploit their past experience and their respective strategy (to decide on the basis of the past experience). Of course the people here are assumed to have all the same informations at any time (and their respective personal experiences) available to decide for themselves independently and parallely; they do not organise among themselves and go to the bar! Had the choice been completely random, the occupation fluctuation of the bar would be Gaussian. Because of the processes involved in deciding to go or not, depending on the respective experience, the occupation fluctuation statistics changes. The "minority" people win such games (all the people "lose" their money if the bar gets "crowded", or more than the threshold, say, 50 here); the bar represents either the "buy" or "sell" room and the (single) stock fluctuations are expected to be represented well by the model. The memory size and the bag of tricks for each agent in the Minority Game model made this process and the resulting analysis very precise [2]. Still, as we mentioned earlier, it cannot explain the origin of the ubiquitous inverse cubic law of fluctuations (see e.g. [3]).

We extend here this simple bar problem to many bars (or from single stock to many), and define the Kolkata Restaurant problem. The number of restaurants in Kolkata, unlike in Santa Fe, are huge. Their (pleasure level) thresholds are also widely distributed. The number of people, who choose among these restaurants, are also huge! Additionally, we assume that the decision making part here in the Kolkata Restaurant problem to be extremely simple (compared to the El Farol bar problem): if there had been any "bad experience" (say, crowd going beyond threshold level) in any restaurant any evening, all those who came there that evening avoid that one for a suitably chosen long time period (T) and starting next evening redistribute this extra crowd or load equally among the rest of the restaurants in Kolkata (equal or democratic crowd sharing). This restaurant or the stock fails (for the next T evenings). As mentioned before, this failure will naturally increase the crowd level in all the other restaurants, thereby inducing the possibility of further failure of the restaurants in service (or service stocks). If T is finite but large, the system of restaurants in Kolkata would actually organise itself into a "critical" state with a robust and precise inverse cubic power law of (occupation or in-service number) fluctuation. This we will show here analytically.

2 Model

Let n represent the number of restaurants in Kolkata. They are certainly not identical in their size and comfort threshold levels. let $p_1, p_2, \ldots, p_n$ denote respectively the crowd threshold levels of these n restaurants. If, in any evening, the crowd level p in the ith restaurant exceeds p_i, then all the p number (fraction, when normalised) of persons coming to the ith restaurant that evening decide not to come to that restaurant for the next T evenings, and the ith restaurant goes out of service for the next T evenings (because others got satisfaction from the restaurants they went last evening, and therefore do not change their choice). If N is the total number (assumed to be large and fixed) of people regularly going to various restaurants in Kolkata, and if we assume that people choose completely randomly among all the restaurants in service (democratic or equal load sharing hypothesis and "knowledge" of the "in-service" restaurants available to everybody), the "avalanches" dynamics

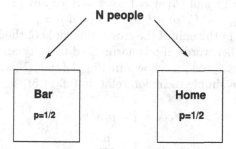

Fig. 1. El Farol bar problem: To go (buy) or not to go (sell) to a single bar (stock). Each of the N people have a choice to stay in their respective homes (collectively represented by a single 'Home' here) or go to the bar each evening. The bar has a pleasure threshold $N/2$ ($p = 1/2$) beyond which people get disappointed (lose the game) and the 'minority' deciding to stay at home that evening win. In the reverse case, the bar people become minority and each of them win.

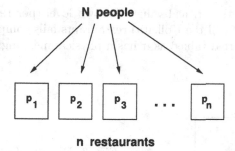

Fig. 2. Kolkata Restaurant problem: Many choice (stock) problem. N people in the city, each has the same choice to go to any of the restaurants (stocks) and therefore, each retaurant gets $p = N/n$ fraction of the crowd to start with. If $p > p_i$ on any evening, then the p fraction of the people going to that ith restaurant gets dissatisfied, and the ith restaurant (stock) falls out of choice for the next T evenings. p then increases to $N/(n - 1)$ and that may lead to a further failure of the jth restaurant (if $p_j < N/(n - 1)$) and so on.

of these restaurants to fall out of service, can be analytically investigated if $T \to \infty$ and the threshold crowd level distribution $\rho(p_i)$ for the restaurants are known (see Figs. 1 and 2).

3 Avalanche Dynamics: Infinite T

This avalanche dynamics can be represented by recursion relations in discrete time steps. Let us define $U_t(p)$ to be the fraction of in-service restaurants in the city that survive after (discrete) time step t (evenings), counted from the time $t = 0$ when the load or crowd (at the level $p = N/n$) is put in the system (time step indicates the number of crowd redistributions). As such, $U_t(p = 0) = 1$ for all t and $U_t(p) = 1$ for $t = 0$ for any p; $U_t(p) = U^*(p) \neq 0$ for $t \to \infty$ if $p < p_c$, and $U_t(p) = 0$ for $t \to \infty$ if $p > p_c$.

If p is measured in the unit of the crowd threshold of the biggest restaurant in Kolkata, or in other words, if p is normalised to unity and $\rho(p_i)$ is assumed to be uniformly distributed as shown in Fig. 3 (and $T \to \infty$ as mentioned), then $U_t(p)$ follows a simple recursion relation (cf. [4,5])

$$U_{t+1} = 1 - p_t; \quad p_t = \frac{N}{U_t n}$$

$$\text{or,} \quad U_{t+1} = 1 - \frac{p}{U_t}. \tag{1}$$

In equilibrium $U_{t+1} = U_t = U^*$ and thus (1) is quadratic in U^* :

$$U^{*^2} - U^* + p = 0.$$

The solution is

$$U^*(i) = \frac{1}{2} \pm (p_c - p)^{1/2}; \quad p_c = \frac{1}{4}.$$

Here $p_c = N_c/n$ is the critical value of crowd level (per restaurant) beyond which the system of (all the Kolkata) restaurants fails completely. The quantity $U^*(p)$ must be real valued as it has a physical meaning: it is the fraction

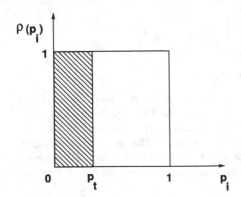

Fig. 3. Density $\rho(p_i)$ of the crowd handling capacities p_i of the restaurants. It is assumed here to be uniform upto a threshold value (normalised to unity). At any evening t, if the crowd level is p_t, restaurants having $p_i \leq p_t$ all fail and the fraction $1 - p_t$ of restaurants remain in service after that evening.

of the restaurants that remains in service under a fixed crowd or load p when the load per restaurant lies in the range $0 \leq p \leq p_c$. Clearly, $U^*(0) = 1$. The solution with $(+)$ sign is therefore the only physical one. Hence, the physical solution can be written as

$$U^*(p) = U^*(p_c) + (p_c - p)^{1/2}; \ U^*(p_c) = \frac{1}{2} \text{ and } p_c = \frac{1}{4}. \qquad (2)$$

For $p > p_c$ we can not get a real-valued fixed point as the dynamics never stops until $U_t = 0$ when the network of restaurants get completely out of business!

3.1 Critical Behavior

At $p < p_c$

It may be noted that the quantity $U^*(p) - U^*(p_c)$ behaves like an order parameter that determines a transition·from a state of partial failure of the system (at $p \leq p_c$) to a state of total failure (at $p > p_c$) :

$$O \equiv U^*(p) - U^*(p_c) = (p_c - p)^\beta; \ \beta = \frac{1}{2}. \qquad (3)$$

To study the dynamics away from criticality ($p \to p_c$ from below), we replace the recursion relation (1) by a differential equation

$$-\frac{dU}{dt} = \frac{U^2 - U + p}{U}.$$

Close to the fixed point we write $U_t(p) = U^*(p) + \epsilon$ (where $\epsilon \to 0$). This gives

$$\epsilon = U_t(p) - U^*(p) \approx \exp(-t/\tau), \qquad (4)$$

where $\tau = \frac{1}{2}\left[\frac{1}{2}(p_c - p)^{-1/2} + 1\right]$. Thus, near the critical point (for jamming transition) we can write

$$\tau \propto (p_c - p)^{-\alpha}; \ \alpha = \frac{1}{2}. \qquad (5)$$

Therefore the relaxation time diverges following a power-law as $p \to p_c$ from below.

One can also consider the breakdown susceptibility χ, defined as the change of $U^*(p)$ due to an infinitesimal increment of the traffic stress p

$$\chi = \left|\frac{dU^*(p)}{dp}\right| = \frac{1}{2}(p_c - p)^{-\gamma}; \gamma = \frac{1}{2}. \qquad (6)$$

Hence the susceptibility diverges as the average crowd level p approaches the critical value $p_c = \frac{1}{4}$.

At $p = p_c$

At the critical point $(p = p_c)$, we observe a different dynamic critical behavior in the relaxation of the failure process. From the recursion relation (1), it can be shown that decay of the fraction $U_t(p_c)$ of restaurants that remain in service at time t follows a simple power-law decay:

$$U_t = \frac{1}{2}\left(1 + \frac{1}{t+1}\right),\tag{7}$$

starting from $U_0 = 1$. For large t $(t \to \infty)$, this reduces to $U_t - 1/2 \propto t^{-\delta}$; $\delta = 1$; a power law, and is a robust characterization of the critical state.

3.2 Universality Class of the Model

The universality class of the model can be checked [4] taking two other types of restaurant capacity distributions $\rho(p)$: (I) linearly increasing density distribution and (II) linearly decreasing density distribution of the crowd fraction thresholds p within the limit 0 and 1. One can show that while p_c changes with different strength distributions $(p_c = \sqrt{4/27}$ for case (I) and $p_c = 4/27$ for case (II), the critical behavior remains unchanged: $\alpha = 1/2 = \beta = \gamma, \delta = 1$ for all these equal crowd or load sharing models.

3.3 Fluctuation or Avalanche Statistics

For $p < p_c$, the avalance size m can be defined as the fraction of restaurants falling out of service for an infinitesimal increase in the global crowd level (c.f. [6])

$$m = \frac{dM}{dp}; \ M = 1 - U^*(p).\tag{8}$$

With $U^*(p)$ taken from (2), we get

$$p_c - p \sim m^{-2}.$$

If we now denote the avalanche size distribution by $P(m)$, then $P(m)\Delta m$ measures Δp along the m versus p curve in (8). In other words [6,7]

$$P(m) = \frac{dp}{dm} \sim m^{-\eta}, \ \eta = 3.\tag{9}$$

4 Avalanche Dynamics: Finite T

The above results are for $T \to \infty$, i.e, when any restaurant fails to satisfy its customers, it falls out of business, and customers never come back to it. This would also be valid if T is greater than the relaxation time τ defined in

(4). However, if such a restaurant again comes back to service (people either forget or hope that it has got better in service and start choosing it again) after T evenings, then several scenerios can emerge.

If T is very small, the recursion relation (1) can be written as

$$U_{t+1} = 1 - \frac{p}{1 - U_{t-T} + U_t}. \tag{10}$$

In fact, if T is large, $U_{t-T} = 1$ and (10) reduces to (1). However, if T is very small, say of the order of unity, then at the fixed point, $U_{t+1} = U^* = U^*_{t-T} = U_t$ and one gets

$$U^* = 1 - p.$$

From (8) one gets

$$P(m) \sim \delta(m - m_0) \tag{11}$$

where m_0 is determined by the initial conditions.

5 Conclusions

We generalise here the El Farol bar problem to that of many bars. This leads us to the Kolkata restaurant problem, where the decision to go to any restaurant or not is much simpler; it still depends on the previous experience of course, as in the El Farol bar problem, but does not explicitly depend on the memory size or the size of the strategy pool. Rather, people getting disappointed with any restaurant on any evening avoids that restaurant for the next T evenings. This generalised problem can be exactly analysed in some limiting cases discussed here. In the $T \to \infty$ limit, the fluctuation in the restaurant service can be shown to have precisely an inverse cubic behavior (see eqn. (9)), as widely seen in the stock market fluctuations. For very small values of T, the fluctuatuation distribution become δ-function like (see eqn. (11)). For large but finite T, the system of restaurants in Kolkata will survive at a (self-organised) critical state [8].

Acknowledgement. I am grateful to Arnab Chatterjee for his criticisms, discussions and help in finalising the manuscript.

References

1. Arthur WB (1994) Am. Econ. Assoc. Papers & Proc. 84: 406.
2. Challet D, Marsili M, Zhang Y-C (2005) Minority Games: Interacting agents in Financial Markets, Oxford Univ. Press, Oxford.
3. Mantegna RN, Stanley HE (1999) An Introduction to Econophysics, Cambridge University Press, Cambridge.

4. Bhattacharyya P, Chakrabarti BK (2006) Eds., Modelling Critical and Catastrophic Phenomena in Geoscience, Springer-Verlag, Heidelberg.
5. Chakrabarti BK (2006) Physica A 372: 162.
6. Pradhan S, Bhattacharyya P, Chakrabarti BK (2002) Phys. Rev. E 66: 016116.
7. Hemmer PC, Pradhan S (2007) Phys. Rev. E 75: 046101.
8. Chatterjee A (unpublished).

Part IV

Comments and Discussions

Comments and Criticisms:
Econophysics and Sociophysics

1 Beyond Econophysics?

Mauro Gallegati

Dipartimento di Economia, Università Politecnica delle Marche, Ancona, Italy
mauro.gallegati@univpm.it

So far econophysics has given contributions in four areas of economics: financial markets, wealth and income distribution, industrial economics (*firms' size distribution, growth rates*) and, more recently, networks analysis. According to Gallegati et al., 2006, there are some weakness in the approach: a lack of awareness of work that has been done within economics itself; resistance to more rigorous and robust statistical methodology; the belief that universal empirical regularities can be found in many areas of economic activity; the theoretical models which are being used to explain empirical phenomena. The paper raised a lively debate (see e.g. Ball, 2006).

In this note I would like to take a different approach: to point out an approach to economics, which goes beyond the *reductionist* approach of the mainstream toward an empirically sound discipline. Economics being a complex system, one should investigate the *emergence* of macro out of the individual behavior and interactions. This opposes the standard view according to which macro is reducible to micro: the work by Gallegati et al., 2007, represents a first step in this direction. The prisoner dilemma's game is a very nice example of the failure of the reductionist hypothesis of the mainstream, according to which if one maximize the satisfaction of one element maximize the satisfaction of the system itself, since some elements can take advantage of interactions.

The usual trick to build macroeconomics on microfoundations is to use the representative agent (RA) framework. Unfortunately, as Hildenbrand and Kirman, 1988, note: 'There are no assumptions on isolated individuals, which will give us the properties of aggregate behavior. We are reduced to making assumptions at the aggregate level, which cannot be justified, by the usual individualistic assumptions. This problem is usually avoided in the macroeconomic literature by assuming that the economy behaves like an individual. Such an assumption cannot be justified in the context of the standard model.'[1]

Econophysicists should help economists to make economics a *falsifiable* discipline not an *axiomatic* one as it is now. The epistemological paradigm of Milton Friedman (1953) is still 'accepted as the correct one within the neoclassical scientific paradigm' (a very lucid presentation in Calafati, 2007). According to Friedman the ultimate goal of a positive science is to develop hypotheses that yield 'valid and meaningful' *predictions* about actual phenomena and which can be *falsified*. Not falsifiable are the underlying hypotheses ('the more significant the theory, the more unrealistic are the assumptions', p. 14) according to a quite old tradition, which dates back to the *aurispici*, supposed to be able to predict the future by interpreting the animals' *interiora* to forecast the future. A bizarre epistemology indeed, but the only one if you wish to use the implausible assumptions of the neoclassical framework (or if you want to 'prove' that Santa Claus really exist).

The mainstream microfoundation literature, which ignores interactions and heterogeneity, has no tools for connecting the micro and the macro levels, beside the RA whose existence is at odds with the empirical evidence (Stoker, 1993) and the general equilibrium theory as well (Kirman, 1992). Also, the standard econometric tools are based upon the assumption of an RA. If the economic system is populated by heterogeneous (non necessarily interacting) agents, then the problem of the microfoundation of macroeconometrics becomes a central topic, since some issues (e.g., cointegration, Granger-causality, impulse-response function of structural VAR) lose their significance (Forni and Lippi, 1997). An Agent Based Modelling strategy has to be seen as a first step toward modeling serious microfoundations.

In a nutshell, we might recall the main differences between the mainstream economics and the Agent Based approach as follows: according to the mainstream we have: perfect information, market clearing, full rationality, no direct interaction (only through prices), while the methodology follows Friedman, 1953 (economists should formulate 'valid and meaningful' *predictions* about actual phenomena regardless the underlying hypotheses). Differently, the group of *Ancona* proposes an approach based upon limited information, out of equilibrium dynamics (path dependency etc.), adaptive behavior, direct interaction and an Agent Based methodology. By it, we mean to model *local interaction* at the single agents level (simple behavioral rules), possibly changing the *interaction nodes* and the *individual rule (adaptation)*. From it, some *statistical regularities* emerge which can not be inferred from the individual behavior (*self emerging regularities*); this *emergent behavior* feeds back

---

[1] See Arrow (1951), Sonnenschein (1972), and Mantel (1976) on the lack of theoretical foundations of the proposition according to which the properties of an aggregate function reflect those of the individual components.

to the individual level (*downward causation*) and each and every proposition may be falsified at micro, meso and macro levels. This approach opposes the axiomatic theory of economics, where the optimization is the rule for a non ad-hoc modeling.

Our hope is that the now mainstream approach based on the RA will be no representative of economics in the very near future.

References

1. Gallegati M, Delli Gatti D, Gaffeo E, Giulioni G, Palestrini A (2007) Emergent Macroeconomics, Springer.
2. Gallegati M, Keen S, Lux T, Ormerod P (2006) Worrying Trends in Econophysics, Physica A 370: 1–6.
3. Ball P (2006) Culture Crash, Nature 44: 686.
4. Arrow K (1951) Social Choice and Individual Values. Wiley & Sons, New York.
5. Forni M, Lippi M (1997) Aggregation and the Microfoundations of Dynamic Macroeconomics. Oxford University Press, Oxford.
6. Hildebrand W, Kirman A (1988) Equilibrium Analysis. North-Holland, Amsterdam.
7. Kirman A (1992) Whom or What Does the Representative Individual represent?, Journal of Economic Perspectives, 6:117–136.
8. Mantel R (1976) Homothetic Preferences and Community Excess Demand Functions, Journal of Economic Theory, 12:197–201.
9. Sonnenschein H (1972) Market Excess Demand Functions, Econometrica, 40:549–563.
10. Stoker T (1993) Empirical Approaches to the Problem of Aggregation Over Individuals, Journal of Economic Literature, 21:1827–1874.
11. Calafati AG (2007) Milton Friedman's Epistemology, Quaderno di ricerca, 280, UPM.
12. Friedman M (1953) Essay in Positive Economics, Chicago University Press.

2 Econophysics and Sociophysics:
Comparison with Earlier Interdisciplinary Developments

Bikas K. Chakrabarti

Theoretical Condensed Matter Physics Division
and Centre for Applied Mathematics and Computational Science,
Saha Institute of Nuclear Physics, 1/AF Bidhannagar, Kolkata 700064, India

Physics, even counted from the time of Isaac Newton (1643–1727), is now about three hundred years' old. It is presently the most matured, successful and dependable of all the natural sciences. It is only natural therefore that along with its own developments, physics had also explored, often very successfully, some other natural science territories and created for example the (not so unconventional any more) branches of physics like astrophysics, bio-physics and geophysics. And econophysics looks like to be a recent addition to this kind of endeavor.

Intense researches on astrophysics and biophysics are now conducted by the physicists in their own departments (not in astronomy or biology depart-ments; although a few 'older' departments are still named 'Department of Physics & Astronomy'!). These research results are also published in regular physics research journals. Nobel prizes in physics have also been awarded to some of these outstanding 'interdisciplinary physics' researchers (at least four astrophysics Nobel prizes so far)! It might be noted in this connection that in astrophysics research, consistent and thorough knowledge development were appreciated from the very beginning (with appropriate emphasis on the scanty observational results available at any stage), and this also percolated the bio-physics research community recently (with the 'old' hang-ups given away) and considerable progresses are being made these days (note, the change-over to the present-day molecular biology, from the cellular one, occurred through the X-ray structure determination of DNA by physicists like Francis Crick and collaborators in the mid-fifties). As such, both these research streams are now very much part of any physics departmental activity, as also of any of the professional physics journal.

Although the main researches in the important area of geosciences are only physical in nature, no regular geophysicists can be commonly found in the physics departments. The same is true for geophysics research papers: they are not regularly published in standard physics journals. To my mind, excessive emphasis on and appreciation of too many disconnected observations, without any attempt to comprehend them, had been the root cause of its failure in inspiring their colleagues like physicists, or for that matter, others. And of course there has been no Nobel prize yet for geophysicists (except perhaps to Edward Appelton for ionosphere research in 1947)!!

Crudely speaking, the main-stream physics research is now composed of two major branches: one looking for the basic constituents of matter and their interactions and mechanics, and the second part deals with the collective (dynamical) properties or behavior of a 'many-body' assembly (typically of the size of Avogadro number of order 10^{23}) of such constituents. After the advent of modern computers in the last thirty years or so, considerable development in understanding these 'collective dynamics' and the consequent 'emergent features' in the dynamics of such many-body systems, especially when each of the constituent follows very simple (local in space and time) but nonlinear dynamics, has taken place. A striking observation in these studies had been the 'self-organised emergence' of 'globally tuned' patterns out of their collective dynamics and their 'universality' classes, independent of the details of the microscopic dynamics of its constituents. Understanding of the 'global' effects of the 'frustrating' constraints among the dynamics of the constituents are now reasonably matured.

All these encouraged the physicists to check and explore their earned knowledge to the well-known many-body systems in the society: like in economics and sociology (mainly in the study of social networks dynamics). Not unlike in the previous attempts and developments, these unconventional applications also try to bring these researches (in econophysics and sociophysics) within the regular (departmental) activities of physics researches! Happily, several main-stream physics journals (like European Physical Journal B, Europhysics Letters, Physica A, Physica D, Physics Letters A, Physical Review E, Physical Review Letters, Journal of Physica A: Mathematical and General, Journal of Physics: Condensed Matter, International J of Modern Physics B, International J of Modern Physics C, and review journals like Physics Reports, Reports on Progress in Physics etc) are regularly publishing research papers in econophysics and sociophysics for the last six-seven years. Like in the initial stages of astrophysics and biophysics, there are some similarities in criticisms from the mainstream economists who essentially tend to ignore these developments in view of their fixed mind-set (of axiomatic foundations most often, and occasionally of 'understanding' the 'natural economic and financial phenomena' claimed by each in terms of their own, but mutually orthogonal, ideas). We believe, however, there are signs of mutual reconciliations emerging. In particular, the balanced emphasis on observations and on developing rigorous analysis of 'toy' models for comprehending only one or some crucial feature(s) of such observations (and not, to start-with, attempt for all the known aspects of the observations), a culture mainly contributed by the econophysicists recently, will help both the streams, physics and economics, in healthy developments.

In short, I believe, criticisms for any such new development are only too natural and have not been uncommon earlier. (Before early sixties, astrophysics was not considered worthy of Nobel prize or for that matter, even for regular funding. Pioneering astrophysicists like Prof. M.N. Saha (1893–1956), our institute's founder, had to undertake projects in 'main-stream' physics of nuclear science in those days to continue their researches!). Anyway, the previous 'successes' with astrophysics (interdisciplinary science of physics and astronomy), biophysics (interdisciplinary science of physics and biology) and the 'not-so-impressive successes' of geophysics (interdisciplinary science of physics and geology) can indeed help us showing the way to succeed with econophysics/sociophysics too. We all should remember, the criticisms die away because the 'old guards' themselves die away and younger researchers come forward with fresh minds! Also, compare the timescales involved in gaining recognition and successes in both astrophysics and biophysics (or for that matter in geophysics)! With the success already in the last ten or fifteen years in starting the econophysics researches in the physics departments of various universities, of having already a set of very knowledgeable researchers and referees in various established physics journals in appreciating good researches in this interdisciplinary field (and also criticising the others), I believe, econophysics and sociophysics researches have already scored a critical mass and are poised to make soon major contributions in science.

3 Creating Principle of Econophysics

Taisei Kaizoji

Division of Social Sciences, International Christian University,
Mitaka, Tokyo 181-8585 Japan
kaizoji@icu.ac.jp

Econophysics has found many empirical laws in regard to the price fluctuation of the financial market through analyzing the high frequency data. Especially, findings of statistical laws of the market such as power law of price fluctuations and multi-fractality of price volatility have a big impact on science as well as the public. Power law of price fluctuation has been known among the researchers from the time before, but economics and the finance have thought that large fluctuations of prices, which is source of power laws, can be ignored and was called 'anomalies', because these phenomena are irrational, and are above one's apprehension. Econophysics makes clear that these laws hold in rather normal states of the financial market, and this has contributed to the development of the technique which analyzes the financial market from a new point of view. For econophysics to contribute to the development of social sciences of the 21st century, it is necessary to establish the principle of the new market mechanism which explains the statistical laws which are discovered so far.

4 Corner and Hope: On The Future Topics of Econophysics

Yougui Wang

Center for Polymer Studies, Department of Physics, Boston University,
Boston, MA 02215, USA and
Department of Systems Science, School of Management, Beijing Normal University,
Beijing, 100875, People's Republic of China

I present some discussions on the changing topics of econophysics. Two trends should be paid more attention, one is to reconstruct economics from perspective of physicists, the other is to enter the domain of economics and find proper topics. Both call for more cooperation between economists and physicists.

Since the term *econophysics* was coined by H.E. Stanley in 1995 in Kolkata, this interdisciplinary subject has undergone development for more than a decade. Many stylized facts with respect to financial markets, income or wealth distribution, firm growth as well as economic fluctuations are discovered empirically. Various features of the complexity of an economy become clear to practitioners with more developed methods or approaches of physics and of complexity research being applied into various issues of economics. Several regular conferences have been frequently undertaken in many places. Some books have been recently published to publicize the subject. Consequently, Econophysics has been approbated in the circle of natural science [1]. Some economists have also viewed it as a heterodox school in economics, or an emerging branch of complex economics [2]. It sounds very successful for a short-lived interdisciplinary subject to have so many achievements. But for majority of scientific researchers, econophysics is still an unknown or strange term. Even in the two original families of econophysics: economics and physics, this 'half-breed' is not accepted as a normal inheritor. Remarkably, some economists who are concerning this field also have shown their worry and made some criticisms about its trend [3].

This worry and criticisms do not come from hollow cavity. Instead, they point out the isolation of econophysics. The current topics of econophysics can be seen at related websites and some review articles [4]. The main prevailing ones include the following three aspects: Stochastic volatility of financial markets, income, wealth, or firm scale distribution, and minority game. The first one led to the birth of econophysics [5]. Its particularity lies in the self-creating of its topic. Application of statistical physics into the abundant data of financial markets generated many stylized facts and then inspired some theoretical investigations to explore the mechanism behind the universe laws. Thus, this topic has no relations to those of mainstream economics. In contrast with this one, the latter two were inspired by works of economists. The issue of wealth distribution is attributed to Vilfredo Pareto, while the problem of minority game to Brian Arthur [7]. However, when physicists tried to uncover the underlying force that causes the corresponding collective phenomena, they almost undertook the task without knowing much more economics, and sometimes the intention has shifted to other themes irrelevant with economics.

This isolation results in the imbalance of composition of practitioners in this subject. Most econophysicists are from physics, only a few from economics. The attention of economists paid to this subject is not proportionable to the efforts that physicists have put into. The zest of economists to attend the meeting of econophysics is less than expected. Papers on econophysics have been published primarily in journals of physics and statistical mechanics, rather than in leading economics journals.

In order to break econophysics away from this corner, more efforts are needed. There must be some more valuable facts that can be discovered from statistics in other aspects, but it will not dominate the field since this sort of data is scarce. The core of topics of this subject should shift to the following two areas. One is reconstruction of economics, which is strongly advocated by Joseph McCauley [1]. The other is entering the domain of economics and finding which is suitable to physicist among existing topics in economics. It must absorb the essence of economics to rebuild it, rather than discard everything. For example, the reconstruction of market theory must be based on the restatement of supply and demand. Amending should be set off from the basic principles of economics, which require sufficient knowledge of economics. Thus even though a few physicists have strong willingness and are competent to carry out this task, it is necessary to cooperate with economists.

Most econophysicists are prepared with advanced tools to solve the selected problems from economics. We believe that more and more approach and methods developed in physics and in complexity research can be applied into various issues in economics. For instance, complex network may replace input-output analysis when we examine the chain of production, combined with economic motivation analysis. The non-equilibrium process analysis will take effect in studying the phenomena of economic growth and development. The precondition for us to successfully apply these methods is perceiving and comprehending the configuration and performance of an economy or any parts in advance. This also needs physicists to go deeply into economics and go along with economists hand in hand.

References

1. 'Econophysicists Matter' (2006) Editorial, Nature 441: 667
2. Barkley Rosser,Jr. J (2006) The Nature and Future of Econophysics, in Chatterjee A, Chakrabarti BK (Eds), Econophysics of Stock and other Markets, pp. 225–234.
3. Gallegati M, Keen S, Lux T, Ormerod P (2006) Physica A 370: 1–6
4. Chakrabarti BK, Chakraborti A, Chatterjee A (2006) Econophysics and Sociophysics : Trends and Perspectives, Wiley-VCH, Berlin.
5. Mantegna RN, Stanley HE (1999) An Introduction to Econophysics: Correlations and Complexity in Finance, Cambridge University Press, Cambridge.
6. Chatterjee A, Yarlagadda S, Chakrabarti BK (2005) Econophysics of Wealth Distributions, Springer-Verlag Italia, Milan.
7. Challet D, Marsili M, Zhang Y-C (2005) Minority Games, Oxford University Press, Oxford.

5 A Poor Man's Thoughts on Economic Networks

Yoshi Fujiwara

NiCT/ATR CIS Applied Network Science Lab, Kyoto 619-0288, Japan
yfujiwar@atr.jp

I make a brief comment on importance of networks in economics.

As the series of this workshop on econophysics have shown, research activities have recently been increased on real economy more than before.

In a nutshell, *heterogeneous interaction* and *aggregation* are focused in their treatment, being reconsidered and reconstructed in ways reported recently by economists [1,2]. Indeed, heterogeneous interactions are taking place as social relationships in which agents are embedded and, at a same time, are forming the relations themselves. Let me recall importance of such economic networks here only briefly.

Real economy has its driving force in production. *Production* refers to a line of economic activities in which firms are putting *added-values* on their products of goods and services. They buy intermediate goods from "upstream" firms, add values on them, and sell products to "downstream" firms or consumers in the end of the line.

Consider a ship manufacturer, for example. The firm buys a number of intermediate goods including steel materials, mechanical and electronic devices, etc. and produces ships. The manufacturer puts added-value on the products, in *anticipation* for return of profits, of course.

In the upstream side of the ship manufacturer is a processed steel manufacturer, which in turn buys intermediate goods like raw steel and fabricating machines. The steel processor may sell its products not only to the single ship company but also to others, or even to other business sectors such as train-vehicle manufacturer. A similar story goes in the downstream side, too.

The entire line of these processes of putting added-values in turn forms a giant network of production ranging from upstream to downstream, down to consumers. Each process of production also requires *labor* and *financing* as inputs. Thus, production, labor and finance are the three corner stones for all economic activities, as written in every textbooks. Even so, little has been hitherto studied, at a nation-wide scale, about the structure and temporal changes in the economic networks of production, financing (firms-banks and inter-banks) and labor (firms-labor).

Why do these matter? Well, in many ways.

At a macroeconomic level, GDP is a net sum of added values in the production network. Demand by agents in downstream positions will be increasing and decreasing in heterogeneous manners; China may increase demand for ships manufactured overseas but not for textile products, for example. The influence of increasing or decreasing demand will propagate along economic networks. And so forth.

To add one thing, let me recall that each firm attempts to put added value in anticipation for return of profits — anticipation, because no firm knows how their produced goods be actually demanded by downstream firms or consumers. Also many firms are facing uncertainty in the change of cost for goods produced in upstream as well as the change of labor and financial costs. Therefore, only *a posteriori*, profits are determined through the interaction between a firm and its upstream and downstream firms.

Each link in such a giant network of production is basically a commercial credit relation in the sense that one firm has only uncertain information about the other's financial state, which causes a risk in payment/receipt and lending/borrowing. Once a firm goes into financial insolvency state, its upstream firms are creditors who are not necessarily able to receive the scheduled payment. Then a creditor has its balance-sheet deteriorated in accumulation, and may eventually go into bankruptcy. This is a chain of bankruptcy.

One may think that such a chain of failure is a rare event and has limited influence. I have recently observed that chain of firms bankruptcy is by no means negligible. Considering the large number of creditors involved in bankruptcy occurred in those cases, due to a "scale-free" character in degree distribution, this has a considerable effect to macroeconomic activity. For instance, bad debts resulting from firms affect the banking system, which would reduce the supply of credit, and could affect an economy-wide shrinkage of credit along the financial network eventually increasing the risk of further bankruptcies.

There are many things to be studied on economic networks, especially on network structures, dynamics and multiple attributes of agents in the networks, how to aggregate the heterogeneous interactions to relate to macroscopic description in economics.

References

1. Delli Gatti D, Gaffeo E, Gallegati M, Giulioni G, Palestrini A (2007) Emergent Macroeconomics: An agent-based approach to business fluctuations. Forthcoming.
2. Aoki M, Yoshikawa H (2007) Reconstructing Macroeconomics: A perspective from statistical physics and combinatorial stochastic processes. Cambridge University Press.

6 Whither Econophysics?

Sitabhra Sinha

The Institute of Mathematical Sciences, C. I. T. Campus, Taramani, Chennai - 600 113, India sitabhra@imsc.res.in

The third of the series of Econophys-Kolkata meetings provides an appropriate forum for reflecting on the prospects for econophysics in the near future. According to several sources, the name "econophysics" was coined at Kolkata itself (or, as it was known at that time, Calcutta) by H E Stanley in 1995, during his talk at a statistical physics meeting in the city. In the decade or so following this emergence of the field as a distinct sub-discipline within physics, we have seen an astonishing increase in the number of papers and people working in this field. For a time, *Physica A*, one of the journals publishing a large fraction of econophysics papers, even ran a virtual journal in the field. There have been popular books, technical monographs, conference proceedings, and even reprint collections. To be sure, there were physicists working on economics-related topics even before the 1990s. This is not even counting people who had moved full-time into economics after being trained as physicists. However, the advent of econophysics in recent times has been a qualitatively different phenomenon, where economists are at last taking notice, even if critically, of the incursion of physicists into their field.

Central to the dialogue between physics and economics at present is the question of whether economic problems are amenable to the kind of analytical skills that physicists are trained in. One can even ask, is economics an empirical science in the same way that physics is. Part of the tense stand-off between the practitioners of the two fields is because of the two conflicting models of what kind of science economics seeks to be. On one hand is the physics-based model of a science based on measurable variables; on the other is the model of mathematics, with a central body of "reasonable" axioms from which all other statements are derived. While Newtonian physics (based on classical mechanics) has indeed served as a model for economists upto the 19th century, in the 20th century the mathematical school has held sway. While physicists might find it strange that entire theoretical edifices are constructed on concepts such as "utility" which cannot be properly measured in any real situation, to many economists the efforts of physicists seem dangerously ad-hoc, often built only to match some empirically observed economic pattern, with little attention being paid to the validity of the assumptions being made in constructing such theories.

One of the central challenges that we as physicists working in economics must overcome is the great divide between these two scientific models for economics. The difference is not just in the body of techniques one uses to approach a problem, but the very fundamental viewpoint one brings to the field. While the former indeed is a reflection of the differences in the training that scientists in the two disciplines receive, the basic assumptions underlying this training often lead to physicists and economists asking completely different questions about the same system. This in itself is not a bad thing, as approaching a problem from different perspectives allow a more complete understanding. However, this does mean that often economists don't see the point that physicists are trying to make (and vice versa, I am sure).

To give an example, let us consider the case of the long tail for income distribution in society. Vilfredo Pareto at the end of the 19th century had made the observation that the distribution of income in a large number of countries, spanning a wide range of economic systems, follow a power-law form (at least for the highest income end) with an exponent of $\simeq 1.5$. This is possibly the first statement claiming an universal scaling relation, long before similar discoveries were made about critical phenomena in physics. However, in the decades following this pioneering work of Pareto, it has received very little attention from economists. Although a few studies have appeared in economics journals which discussed the validity of Pareto's observation, and Herbert Simon had proposed a model to explain the origin of such scaling behavior, economists by and large did not consider the issue interesting enough to verify it with recent data that were far more accurate than what was available in Pareto's time. It was left to physicists who, starting in the early 1990s, did a series of careful empirical observations from tax records to accurately describe the income distribution for various countries. This work has in its turn led to a number of increasingly sophisticated models to describe how such distributions may arise. Unfortunately, this has not had much of an impact in economics, partly because of the "great divide" between the two schools that I mentioned above.

Of course, there are reasons behind the scepticism with which most economists view econophysics. Physicists do have a reputation for being dilettantes, and some papers published in the field show the authors to be ignorant of basic economic concepts, often as fundamental as the distinction between income and wealth. This is analogous to an economist writing a paper in physics without understanding the difference between current and voltage. Several papers in econophysics also seem to be rather strained efforts in that a physics problem has been forcibly molded into economic terms, without proper justification. However, this is a problem common to any young academic discipline, where the standards have not yet been formalized. If physicists consistently keep working on economic problems, while at the same time learning from the large body of knowledge that economists have already acquired, I am sure that eventually economists will be convinced to look at econophysics not as an upstart pretender but as a valuable ally. Physicists will also benefit from learning some of the techniques used regularly by econometricians. For example, often we show the evidence for power law simply by drawing a straight line using least square fit over the data points in a double logarithmic plot. This is however rather unconvincing to social scientists, who have over the years developed quite sophisticated statistical tools (such as, maximum likelihood estimation) for analyzing functional relationships in the data. The flow of such techniques into the physicists' toolbag is one way in which physics can benefit from its dialogue with economics.

Coming to the specific topic of the Econophys-Kolkata meetings, while they have indeed been very successful in terms of generating intellectual ferment, we do need to think about how to take it forward in the future. A momentum has been generated in this country as a result of these meetings, by which a number of young physicists have gotten interested in this topic, and we must make sure of not losing it. While the initial enthusiasm seems to have decreased somewhat, this is the time to consolidate the core group who can spearhead the econophysics movement in India. One point of concern is that although the organizers have tried to bring local economists in the meeting, by and large their participation had been rather lukewarm. This may of course be partly because of the very novelty of the field, and the usual problem of interacting across very different disciplines. However, this is an important issue on which future meeting organizers must work at. Another point worth noting for the future is how to get undergraduate and graduate students in both physics and economics interested in the field, before their training has forced them into the traditional mental straitjackets of the respective disciplines. A possible solution is to have half-day "popular sessions" in future Econophys-Kolkata meetings, where students from both physics and economics can listen to a few talks intended for non-specialists and generally interact with practicing econophysicists so that they can get to know about the subject.

In the longer term, if it is successful, econophysics may have to decide whether it will remain a branch of physics, or integrate with economics. The latter alternative will involve increasing interactions and collaborations with traditional economists, and publishing in prestigious economic journals. Moreover, as economists are more used to communicating their ideas through books, physicists must try to get their message across to economists by means of books, preferably written in collaboration with economists sympathetic to the econophysics venture. Wider recognition will have to wait till the field gets a Noble prize or two. This will of course depend upon whether econophysics has a significant impact on the problems that are perceived to be the outstanding issues of traditional economics. Working out such a common ground between the two fields seems to be the immediate task for us in the near future.

Acknowledgement. I would like to take this opportunity to thank the organizers of the three Econophys-Kolkata Workshops, Arnab Chatterjee and Bikas K. Chakrabarti, for their efforts in making this series of meetings a great success, both in terms of increasing the awareness of the field in the Indian academic community and providing a wonderful forum for physicists and economists, interested in working together, to meet and discuss. I would additionally like to thank Bikas for getting me actively involved in econophysics research and also my teacher and mentor, Pinaki Mitra, who first got me interested in economics.

7 Thoughts on Income and Wealth Distributions

Peter Richmond

School of Physics, Trinity College, Dublin, Ireland, and
School of Computational Sciences, University of East Anglia, Norwich, UK

This meeting was the third in the ECONOPHYS-KOLKATA series covering new development in Econophysics and Sociophysics. The first in 2005 focused on income and wealth distributions. At this meeting a number of speakers expanded on this issue. It is interesting to observe that the first person to study this topic in a quantitative manner, Pareto, was trained as an engineer. In recent years, it is the physics community who have made significant contributions to the topic, again by focussing not only on theoretical methodologies but also making comparisons of their results with empirical data. What is surprising is that whilst economists have advised governments on income policy over the years, few in the economics community have expressed much interest in the fundamentals of the subject during the 100 years or so since Pareto made his first observations.

What does seem clear from the mounting evidence is that income and wealth distributions across societies everywhere follow a robust pattern and that the application of ideas from statistical physics can provide understanding that complements the observed data. The distribution rises from very low values for low income earners to a peak before falling away again at high incomes. For very high incomes it takes the form of a power law as first noted by Pareto. The distribution is certainly not uniform. Many people are poor and few are rich.

However looking at the data more closely reveals a number of subtleties that have not been accounted for in the present models. In Figure 1 we show data for the cumulative distribution of incomes in the UK for the year 1995.

The upper curve is calculated from survey data and tends to a power law which was confirmed by data obtained by Cranshaw [1] from the UK Revenue Commissioners. The slight shift in the two curves is due to uncertainty in a normalisation factor but the power law is clearly seen and extends from weekly incomes of just under £1000 per week up to around £30,000 per week. Over this region the exponent of the power law is ~ -3.3. This might be assumed to be the end of the story with the power law being associated with Pareto's law. However from data published by Forbes and the Sunday Times for the wealth of billionaires, we can make an estimate of the income generated by the wealth. This yields a second power law with exponent ~ -1.26 and this might be identified with the Pareto power law. This suggests what many people believe to be true, namely that the super wealthy pay less tax as a proportion of their income than the majority of earners in society!

To account for this data, a number of models have been proposed. One class studied by Solomon and Richmond, that might be considered to constitute a mesoscopic approach is based on a generalized Lotka Volterra models [2–4]. Other microscopic models invoke agents that exchange money via pair trans-actions according to a specific exchange rule. The results from these latter models depend critically on the nature of the exchange process. Two quite different exchange processes have been postulated. The first by Chakrabarti and colleagues [5,6] conserves money during the exchange and allows savings that can be a fixed and equal percentage of the initial money held by each agent. This yields the Boltzmann distribution. Allowing the saving percent-age to take on a random character then introduces a power law character to the distribution for high incomes [7]. The value of the power law exponent however can be shown to be exactly unity [8]. Other modifications are needed to rescue this particular exchange rule.

On the other hand, the model of Slanina [9] assumes a different exchange rule. In addition it allows creation of money during each exchange process and the solution is not stationary. One must normalise the money held by an agent with the mean value of money within the system at time, t. In this

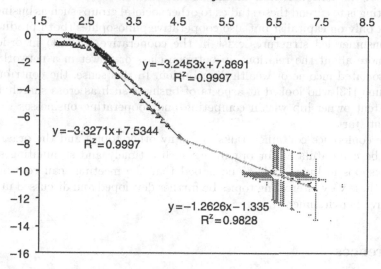

Fig. 1. Log log plot of cumulative distribution of income in the UK for 1995. See text for explanatory detail.

way a stationary solution for the distribution of money can be obtained. Such a procedure must also be done to obtain a solution from the Lotka Volterra approach and it is interesting to see that the final results for both methods yield a distribution function of the same form. Detailed numerical comparisons with the data suggest that this form gives a good fit to the data below the super rich region [10]. To fit the super rich region probably is possible by assuming a non linear functional form for the agent exchange parameter that varies inversely with income of the individual.

Further investigation of the UK data shows that the low end power law changes progressively from 3.3 to values just below 3 over the period 1992 to 2002. Further studies over the period 2002 through to the present day would be interesting and perhaps reveal how changes in taxation policy of the UK government affected this exponent over the entire 15 year period. Similar studies across other societies would be equally interesting.

The link between these facets to the structure of a society was highlighted in the meeting by an interesting contribution from Sinha [11] who examined using data from historical sources. The Doomsday Book provides an account of the structure of society in feudal England during the reign of William the first and reveals how power was distributed from the King through to Barons and lesser members of the nobility down to the vassals. Similar studies were presented based on data for military powers and criminal gangs in the US. The situation in Russia would be interesting to study were data available. According to BBC reporter Rupert Wingfield-Hayes [12] fifteen years ago all the wealth in Russia was in the hands of the state; at present 25% of the wealth of the nation is now in the hands of 36 men. What would also be interesting is to extend these studies to other societal groups such as businesses run not only on capitalist but also cooperative philosophies (both production and consumer led structures exist in the cooperative context) in order to learn more about the relationship between class or power in a network and the associated income or wealth distribution. In this sense, the contribution by Souma [13] who looked at aspects of business such as cross shareholding and patent ownership within competing and cooperating businesses was an excellent start.

The conference contained talks on many other topics and the organisers are to be congratulated for organising such a timely and stimulating series of workshops in India. It is to be hoped that the meetings can continue to enable these very interesting topics be further developed and discussed in such a convivial environment.

References

1. Cranshow T (2001) Unpublished poster presentation at APFA 3 London.
2. Solomon S, Richmond P (2001) Int J Mod Phys C 12(3): 1–11
3. Malcai O, Biham O, Richmond P, Solomon S (2002) Phys Rev E 66: 31–102
4. Solomon S, Richmond P (2002) Eur J Phys B 27(2): 257–261
5. Chakraborti A, Chakrabarti BK (2000) Eur Phys J B 17: 167
6. Patriarca M, Chakraborti A, Kaski K (2004) Phys Rev E 70: 016104
7. Chatterjee A, Chakrabarti BK, Manna SS (2004) Physica A 335: 155
8. Repetowicz P, Richmond P, Hutzler S (2005) Physica A 356: 641
9. Slanina F (2004) Phys Rev E 69: 46102
10. Richmond P, Hutzler S, Coelho R, Repetowicz P (2006) in 'Econophysics and Sociophysics: Trends and Perspectives', Eds Chakrabarti BK, Chakraborti A, Chatterjee A, Wiley-VCH, Berlin, chapter 5
11. Sinha S, presented at Econophysics-Kolkata III
12. http://news.bbc.co.uk/1/hi/programmes/from_our_own_correspondent/ 6577129.stm
13. Souma W, presented at Econophysics-Kolkata III

8 Comments on the Econophys-Kolkata III Workshop

Jürgen Mimkes

Department Physik, Universität Paderborn, Warburgerstr. 100, Germany

Econophysics and its conferences have a long tradition in Kolkata. Accordingly, the conference has again been organized in professional manner. The hospitality and friendly environment of the conference, the good contacts between scientists, the fast production of the conference proceedings are most helpful for the attraction of the conference site at the Saha Institute in Kolkata. Regular meetings in Kolkata have several advantages:

1. Regular meetings establish a reliable relationship of the international community with the Kolkata group.
2. Regular conferences in Kolkata give international visitors a chance to feel at home in Kolkata and plan and visit this area and other regions of India after the conference.

However, a single conference site over many years may also have some disadvantages:

1. The long distance from the USA, EU, China or Japan makes it difficult for the international community to attend every meeting, especially, since the number of conferences in this field has grown very much (COST P10, WEHIA, NEW).
2. Most international conferences (on Econophysics) have new conference sites at every meeting. This is a more democratic way for most participants, so every country will have the chance to present its values.
3. Changing sites also raises curiosity of the participants to see new places and meet new people outside of the conference. Especially people working on social science problems may want to have a chance to experience different groups of people in different countries.

My personal opinion on future conferences on Econophysics in Kolkata: The new aspect of the field of econophysics may perhaps be fading, but the scientific interest is still growing. And we have only started to introduce natural science into economics. If thermodynamics would be considered as one of the basic theories of econophysics, I could imagine a large program in the next years:

1. The theory of economics could be expanded to slow changing markets like state economies and companies according to isothermal changes. This would lead to an advanced theory of economic growth.
2. Fast changing processes like financial markets could be investigated according to adiabatic changes. The adiabatic law: $VT^{f/2} =$ constant indicates the same exponential qualities as financial power laws $\log f = A/x^{-m/2}$.
3. Socio-economics is in many ways similar to chemistry. The large number of groups of people relates to the different elements and their compounds. There is a huge amount of work ahead. We have thousands of papers on the solid and liquid state, but few on the collective and individual state of people. We know much about hydrochloric acid, H-CL or carbon monoxide C-O. But how much do we know about Hindu-Muslim relations or Christian-Muslim interactions, about US-India or India-Pakistan?

In order to keep a permanent annual international conference on socio economics in India, I would suggest to make these at least alternating in Kolkata and Chennai to attract international interest. If this does not draw enough international scientists, perhaps more of the many wonderful places of India may have to be included.

In some countries like Romania, national (and some international) interest of economics and physics students has been drawn by summer schools of a few experts in the fields of complexity, econophysics and sociophysics. The summer schools also give the opportunity to establish a wider scope and structure of the fields and give students the opportunity to participate in exercises, problem solving and practical work.

New Economic Windows

Massimo Salzano, Alan Kirman (Eds.)
Economics: Complex Windows
2005, XX, 217 p., Hardcover
ISBN: 978-88-470-0279-1

Arnab Chatterjee, Sudhakar Yarlagadda, Bikas K. Chakrabarti (Eds.)
Econophysics of Wealth Distributions
Econophys-Kolkata I
2005, X, 248 p., Hardcover
ISBN: 978-88-470-0329-3

Arnab Chatterjee, Bikas K. Chakrabarti (Eds.)
Econophysics of Stock and other Markets
Proceedings of the Econophys-Kolkata II
2006, XIV, 253 p., Hardcover
ISBN: 978-88-470-0501-3

Massimo Salzano, David Colander (Eds.)
Complexity Hints for Economic Policy
2007, XXIV, 312 p., Hardcover
ISBN: 978-88-470-0533-4

Arnab Chatterjee, Bikas K. Chakrabarti (Eds.)
Econophysics of Markets and Business Networks
2007, XII, 268 p.,Hardcover
ISBN: 978-88-470-0664-5